Beyond Earth Day

"On April 22, 1970, Americans came together for the very first Earth Day. They came together to make it clear that dirty air, poison water, spoiled land were simply unacceptable. They came together to say that preserving our natural heritage for our children is a national value. And they came together, more than anything else, because of one American—Gaylord Nelson. His career as Wisconsin's Governor, United States Senator, and later as counselor of the Wilderness Society has been marked by integrity, civility and vision.

As the father of Earth Day, he is the grandfather of all that grew out of that event—the Environmental Protection Act, the Clean Air Act, the Clean Water Act, the Safe Drinking Water Act. He also set a standard for people in public service to care about the environment and to try to do something about it."

—President Bill Clinton

Beyond Earth Day

Fulfilling the Promise

Gaylord Nelson

with
Susan Campbell and Paul Wozniak

Foreword by Robert F. Kennedy Jr.
and a preface by Tia Nelson

THE UNIVERSITY OF WISCONSIN PRESS

The University of Wisconsin Press
1930 Monroe Street, 3rd Floor
Madison, Wisconsin 53711-2059
uwpress.wisc.edu

3 Henrietta Street
London WC2E 8LU, England
eurospanbookstore.com

Printed in the United States of America

Library of Congress Cataloging-in-Publication Data
Nelson, Gaylord, 1916–
Beyond Earth Day : fulfilling the promise / Gaylord Nelson with Susan Campbell and
Paul Wozniak; with a foreword by Robert F. Kennedy Jr.
p. cm.
Includes bibliographical references and index.
ISBN 0-229-18040-9 (cloth: alk. paper)
1. Earth Day. 2. Environmentalism. I. Campbell, Susan, 1965– ii. Wozniak, Paul A.
III. Title.
GE195 .N45 2002
333.7'2—dc21 2002002806

ISBN 978-0-299-18044-7 (pbk.: alk. paper)
ISBN 978-0-299-18043-0 (e-book)

A land ethic, then, reflects the existence of an ecological conscience, and this in turn reflects a conviction of individual responsibility for the health of the land. . . .

Quit thinking about decent land-use as solely an economic problem. Examine each question in terms of what is ethically and esthetically right, as well as what is economically expedient. A thing is right when it tends to preserve the integrity, stability, and beauty of the biotic community. It is wrong when it tends otherwise.

Aldo Leopold, "The Land Ethic," *A Sand County Almanac*

Contents

Illustrations

Foreword

Robert F. Kennedy Jr.

I remember what America was like before Earth Day; the Cuyahoga River burned for a week with flames five stories high; Lake Erie was declared dead. As a boy, I was warned not to swim in the Hudson, the Potomac, or the Charles Rivers. I recall how black smoke billowed from the stacks in Washington, D.C., so that we had to dust daily for soot, and how on some days, you could not see the length of a city block. I remember in 1963 when the eastern anatum peregrine falcon—arguably America's most spectacular predatory bird—went extinct, poisoned out of existence by DDT.

On Earth Day in 1970, an accumulation of such insults and the leadership of Senator Gaylord Nelson drove twenty million Americans—10 percent of our population—into the street in the largest demonstration in American history. They demanded that our political leaders return to our people the ancient environmental rights taken from our citizens during the previous eighty years. Motivated by that extraordinary show of grassroots power, Republicans and Democrats working together passed twenty-eight major laws over the next ten years, to protect our air, water, endangered species, wetlands, and food supply. Those laws, in turn, became the model for over 150 nations that had their own versions of Earth Day and started making their own investments in their environmental infrastructure.

The far right complained that Nelson's new environmental movement was a communist plot and pointed for evidence that Earth Day was also Vladimir Lenin's birthday. They accused Nelson of being a radical, divorced from American values. They could not have been more wrong; Gaylord Nelson and his followers were engaged in a battle to reclaim America's most basic values. The environmental movement embodies the essential American principles, including democracy, free market capitalism, property rights, and the central importance of human dignity.

Federal environmental laws democratized our communities in a remarkable way. America's progressive social movements, the women's movement, the labor movement, and the civil rights movement were all intended to make our society more democratic by spreading out power and by allowing our most humble, vulnerable, and alienated citizens to participate in the dialogues that determine the destinies of our communities. I would argue that none of them have accomplished these goals as effectively as has the environmental movement.

Nowadays, we often hear the anti-environmental right wing vow to rid our nation of its federal environmental laws in the name of democracy. "We are going to get rid of the big federal government, and return control to the states," they say. "After all, local control is the essence of democracy, and the states are in the best position to defend, protect, and understand their own environment." The real outcome of that devolution would not be community control but corporate control. Gaylord Nelson understood that federal laws were necessary to protect small communities and individuals because large corporations can so easily dominate the state political landscapes.

The Hudson Valley is only one of thousands of communities across America still struggling with the legacy of the pre–Earth Day version of "community control." The General Electric Company came into the impoverished upstate New York towns of Hudson Falls and Fort Edwards, promising the towns' fathers a new factory, fifteen hundred new jobs, and a raised tax base. All the town had to do was to persuade the state of New York to write GE a permit to allow the company to dump toxic PCBs into the Hudson River. If they refused to cooperate, GE threatened to move its jobs to New Jersey.

Hudson Falls took the bait. A few decades later, GE closed the doors on these factories and left a two-billion-dollar cleanup bill that nobody in the Hudson Valley can afford.

Federal environmental laws were meant to put an end to that kind of corporate blackmail and to stop corporations from whipsawing one community against another in a race to the bottom to lower their environmental standards and recruit dirty industries in exchange for the promise of a few years of pollution-based prosperity.

Today, thanks to Gaylord Nelson and the federal environmental laws Earth Day inspired, communities have the clout to say "no" to companies like GE. If you are an American in almost any state, and some big company tries to put a landfill or an incinerator in your backyard, and you think it is going to destroy your community, you have a good shot at stopping them. You have the right to demand a full environmental impact statement in which the proponent has to disclose all the costs and benefits of that project over time to your community. You have a right to a hearing on that environmental impact statement in which you can cross-examine witnesses and bring in your own witnesses; you have a right to a written transcript of that hearing and to a judicial decision based on a rational interpretation of that transcript. If you do not like the decision, you can appeal.

If there is a polluter in your backyard, and the government fails to enforce the law, you can step into the shoes of the United States Attorney and drag that polluter in front of a federal judge for penalties of $25,000 a day and injunctive relief. Thanks to Gaylord Nelson, we have toxic inventory laws and right-to-know laws that make government and industry more transparent on the local level and force them to disclose how their activities may harm our local communities.

The far right and some industrialists hate the federal environmental laws, which they consider costly and time consuming. In fact, democracy, in the short term, is expensive. It is awkward, painful, and unwieldy, but in the long term there is no system more efficient. The American experiment with civilian nuclear power, which *Forbes Magazine* recently called "the greatest economic catastrophe in the history of mankind," is testimony to that fact. The nuclear industry got its start in the 1950s and 1960s before we had environmental laws that would have allowed the public to scrutinize deceptive industry claims about the economic viability of atomic power. If those laws had been in place, America would have avoided wasting a half-trillion dollars in investment that generations of Americans will be paying for.

A lot of people on Capitol Hill and a lot of the political leaders complain that environmental regulations somehow impede free-market economy. But you might as well claim that laws forbidding piracy or theft impede the free market. The whole point of environmental laws is to impose a true free-market economy by policing and punishing the cheaters.

You show me pollution, and I will show you people who are not paying their own way, people who are stealing from the public, people who are getting the public to pay their costs of production. All environmental pollution is a subsidy. When General Electric dumped PCBs, it was avoiding one of the costs of bringing its product to market—the cost of properly disposing of a dangerous chemical. In avoiding that cost, the company was able to outcompete its competitors and to enrich its shareholders and board, but the cost did not disappear. It went into the fish, and it put the employees out of work; it made the people sick, and it took the land off the tax rolls, and it dried up the barge traffic. Pollution's impacts impose costs on the rest of us that, in a true free-market system, would have been reflected in the price of the polluter's product when it came to market. Like all polluters, GE used chemical ingenuity and political clout to escape the discipline of the free market, outcompete its competitors, and force the public to subsidize its profits.

As an environmental advocate, I use the laws that Gaylord Nelson inspired to reimpose the free market, to get people to pay their own way. That is what all federal environmental laws are about. They force companies to internalize their costs the same way they internalize their profits.

The right wing also argues that environmental laws harm property rights. But if you examine those arguments, it is inescapable that their real objective is to give constitutional protection to the right to pollute. In fact, environmental laws are designed to protect public and private property from polluters. In any case, there is no "right" to use your property in a way that will injure your neighbor's property or such public properties as air, water, and fisheries. The public, after all, owns these resources. The law says that the people of the United States own these "public trust" resources: the running waters, the wandering animals, the fisheries, the shorelines, the wetlands, and the air that we breathe. America's modern environmental rules descend from ancient laws that held that those assets not susceptible to private ownership—the commons—belonged to all the people. These protections date back to the Code of Justinian of Roman times. Under that code, the emperor himself could not sell monopolies over the fish or stop a citizen of Rome from fishing or accessing a shoreline or river. Everyone has a right to use the public trust assets, but never in a way that could diminish or injure their use and enjoyment by other people. These rights were restated in the Magna Carta and descended to the people of the United States following the American Revolution. Rather than imposing new laws as the right wing claimed, Nelson's movement was simply codifying ancient rights that were essential to American democracy and designed to protect public property.

Environmental laws also safeguard that essential American value of community—the idea that we can't advance ourselves as a nation by leaving our poorer brothers and sisters behind or by abandoning our obligations to future generations. Industrial and political leaders have short horizons—the next election or the next shareholder meeting. Their impulse is to treat the planet as if it were a business in liquidation, convert our natural resources to quick cash, and enjoy a few years of pollution-based prosperity. In following this course, we can create an instantaneous cash flow and the illusion of a prosperous economy, but our children are going to pay for this joy ride with denuded landscapes and huge cleanup costs that they simply will not be able to afford and that will amplify over time. Environmental injury is deficit spending that abandons our obligation to pass to the next generation of Americans communities that have the same opportunities for dignity and enrichment as those our parents passed to us.

The people who get hurt most by environmental neglect are the poorest people in our societies. Pollution robs the next generation of their use of the public trust resources, which are their greatest assets: fresh air, clean water, abundant fisheries, pure aquifers, and enriching landscapes to share with their children. The best measure of how our democracy is working is how well it distributes the goods of the land. Pollution is an attack on the poor. This is always true in every environmental controversy. The people who shoulder the heaviest burden of environmental injury are always the poorest people in our society.

Finally, the right wing accuses environmentalists of putting nature before people. The opposite is true. Gaylord Nelson understood that we are not protecting nature for nature's sake. We are protecting nature because it enriches humanity. Nature enriches us economically. The economy is a wholly owned subsidiary of the environment. It enriches us culturally, recreationally, aesthetically, spiritually, and historically. It connects us to one another, to our history and our culture, and to common experiences that give us identity as a people. Human beings have other appetites besides money, and if we do not feed them we are not going to grow and develop. We are not going to become the kinds of beings that we are supposed to become. When we destroy nature we diminish ourselves and impoverish our children.

We are not fighting to save those ancient forests in British Columbia or in the Pacific Northwest, as Rush Limbaugh loves to claim, for the sake of a tree or a spotted owl. We fight because we believe that those trees have more value to humanity when standing than if we cut them down. I am not fighting for the Hudson River for the shad or the salmon, for the striped bass or the sturgeon, but because I believe my life, my children, and my community will be richer if we live in a world where there are shad and sturgeon and stripers and where our children can see the traditional fishermen in their small boats engaged as they have been for generations. That experience connects them to 350 years of New York State history and helps them understand that they are part of something larger than themselves, a continuum, a community. The experience also connects them to ten thousand generations of humankind who lived by nature's bounty, and to God. I do not believe that nature is God, and I don't think that we ought to be worshiping it as God, but I do think that nature is the way that God communicates with us most forcibly. God talks to human beings through many vectors: through each other; through their community; through organized religion, and the great books of those religions; through wise people; through art, literature and poetry, and music; but nowhere with such force, clarity, and detail, and with such texture, grace, and joy, as through creation. For me, therefore, destroying these things is the moral equivalent of tearing the pages out of the last Bible, Torah, or Koran on Earth. It is a cost that I do not think is prudent to impose on ourselves, and I doubt if we have the right to impose it upon our children.

That really is what environmental advocacy is all about—recognizing that we have an obligation to the next generation, and that we live in a community. That part of being a community is painful because we cannot just make decisions based on our own self-interest; we have to take into account the impacts of those decisions on others, and particularly on those members of our community who do not participate in the political process by which these public trust assets are allocated, because they are not born yet, or because they come from neighborhoods that are too poor or dysfunctional. Nevertheless, even if you do not participate, you have a right to breathe clean air and to fish and to swim in our lakes, and your children have those same rights. Those are the rights that Gaylord Nelson's Earth Day retained for the American people.

Preface

Tia Nelson

The greatest gift my father gave me is the fundamental conviction that what we do matters. He didn't just believe this to be true of himself, or me, or a small circle of friends and allies. He believed that in ways both large and small, through both words and actions, each and every person holds the power to change the world in unimaginable ways.

Certainly Papa's life is a great testament to this principle. This small-town boy from Clear Lake, Wisconsin, grew up during the Depression, with the natural world for a playground. At the age of ten, atop his father's shoulders, he listened to Robert La Follette Jr. give a whistle-stop speech in Amery, Wisconsin. That experience inspired him to eventually run for public office and dedicate his life to public service. Papa spoke out against the war in Vietnam, became a consumer advocate, battled big drug companies, and helped pass countless laws to protect and conserve our lands and waters. Yet in spite of his many successes, the story of Earth Day begins as a tale of utter failure.

My father was highly respected for championing the issues of conservation, public lands, and environmental quality during his time as Wisconsin's governor. But in 1960s America, civil rights and foreign policy dominated the national discourse. Papa was determined to raise awareness of the plight of our environment among our nation's top leaders. He would later serve with Robert Kennedy in the Senate; in 1962 Robert Kennedy was attorney general and his brother John was president. Papa reached out to Robert, and the president agreed to a national conservation tour.

But even with a popular president at its helm, the tour failed to galvanize the nation. The president's heart wasn't in it, and press coverage was lackluster at best. The tour began in Pennsylvania in September 1963 and crossed the country, but it didn't get the publicity Papa had hoped for.

Deeply disappointed, but undaunted, Papa spent the next seven years considering how else to thrust the issue of the environment into the political mainstream. As he wrote, "The evidence of environmental deterioration was all around us, and everyone noticed except the political establishment. The environmental issue simply was not to be found on the nation's political agenda. The people were concerned, but the politicians were not."

And then in 1969, not long after touring the oil spill devastation on the coast of Santa Barbara, he read an article describing the impact of campus teach-ins on public opinion about the Vietnam War. "That's it!" he thought to himself, "I'll call for a national teach-in on the environment in schools across the country." His goal was simple: to "shake the political establishment out of its lethargy and force the environmental issue onto the national political agenda."

Two decades before widespread use of the Internet and cell phones, and more than three decades before Facebook, my father, a couple of dedicated Senate staffers, and a small band of seasoned organizers launched the teach-in idea on a shoestring budget. What they envisioned was a single day set aside for teaching about the environment on 2,000 college campuses and in 10,000 elementary, middle, and high schools across the country. What they sparked was a movement. In 1970, an estimated 20 million Americans participated in the first Earth Day. It was the largest secular event in American history.

"Earth Day worked because of the spontaneous response at the grassroots level," my father noted. Earth Day was a resounding success because the organizers didn't try to shape a uniform national action. They empowered ordinary people to express their passion for the Earth in whatever way they chose from wherever they were. So millions did, and politicians took note.

What followed was the "environmental decade," in which more laws were passed to preserve our natural heritage than in the entire rest of our nation's history. The first Earth Day led to the creation of the U.S. Environmental Protection Agency and passage of the Clean Air, Clean Water, and Endangered Species Acts. It was a moment of rare political alignment that elicited support from Republicans and Democrats, rich and poor, city slickers and farmers.

"After Earth Day, nothing was the same," wrote environmental writer Philip Shabecoff in his book *A Fierce Green Fire: The American Environmental Movement*. According to Shabecoff, Earth Day brought revolutionary change and "touched off a great burst of activism that profoundly affected the nation's laws, economy, corporations, farms, politics, science, education, religion, and journalism. . . . Most important, the social forces unleashed after Earth Day changed, at least for a time, the way Americans think about the environment."

Never in Papa's wildest dreams could he have imagined precipitating these events. Never could he have imagined that a day dedicated to the environment would inspire millions to action and alter the course of history. Never could he have imagined the enduring legacy of Earth Day, celebrated 40 years later by nearly a billion people in more than 180 countries.

Yet he would have recognized the important lesson in this story: what we do matters. And while we can hope our actions will make a difference, and try to imagine what impact our words will have, we can never know in advance which of our wild ideas will catch on, which ones will reach into another's heart to spur action, or which ones will leave the world a better place in the end.

Papa worked every day until only months before his death. He never stopped believing that he could make a difference if he just got up, went to work, and applied himself with purpose and integrity to the cause that defined his life. When asked why he still worked at age eighty-nine, he replied, "The job's not done yet."

The job's *still* not done. *Our* job's still not done. How will you use your power to change the world?

February 2012

Introduction

My purpose in writing this book is simple. It is to pick up where we left off thirty years ago, when that first Earth Day served as a national call to arms for the environment.

Where our planet's health is concerned, I have always believed that a public armed with knowledge is a public armed with the means and the determination to find a solution. The first Earth Day demonstrated that when the public understands the urgency of addressing the planet's ecological ills, it acts.

I have faith it will act again.

Sometimes we need a reminder, however. This is a reminder that the work on which we embarked in 1970 is not finished. Far from it. Many of the problems we faced as a nation back then remain with us today. Some have worsened. The intent of this book is to sketch a brief primer of some of the most serious environmental threats with which we as a nation are grappling today.

Those who closely follow environmental issues and developments in the natural sciences are unlikely to find scientific revelations on these pages. Nor will readers find an exhaustive review of the research behind many of today's most pressing environmental problems. That is not the purpose, though I have sought to include the latest studies where appropriate.

Rather, this book is written in the same spirit on which Earth Day was founded: to call the public's attention once again to our most urgent environmental challenges and reawaken in it the same sense of urgency that propelled the modern environmental movement.

The book is divided into four sections. The first, "The Earth and Its Day," examines in chapter 1 the forces that shaped that first Earth Day and presents in chapter 2 a report card on the health of the planet today.

In part 2, "Imperiled Planet," chapters 3 and 4 outline in greater, yet brief, detail the most serious of the environmental problems and their implications for the United States and the world. Chapter 5 characterizes the changing nature of pollution, discussing the influx of synthetic chemicals and chemical byproducts that now blanket the globe—and what scientists are learning about the risks they pose for humans and nature.

Part 3, "Environmentalism: Then and Now," revisits the environmental movement, assessing its accomplishments and its shortcomings. Chapter 6 looks at the environmental movement today and asks whether it is strong enough to tackle the problems we now face. In evaluating the movement, experts attempt to piece together why public fervor for action on environmental issues is lacking at the same time that overall public sympathy for environmental concerns is statistically as strong as ever.

It is one thing to bemoan the unfortunate state of environmental affairs today and quite another to try to do something about it when many say it is hopeless. In part 4, I propose action on several fronts. "An Environmental Agenda for the Twenty-first Century" is an extension of some of the same principles I have espoused for the last thirty years, with some new twists for a new age.

Some of what is said here will be controversial and is apt to rankle quite a few. That's happened before, as anyone who is an environmentalist knows. However, if what is said or proposed provokes action and discussion, it is well worth the effort.

PART 1
The Earth and Its Day

1

Earth Day

When the People Spoke

If we have the will, the environmental challenge can be met.
—1970

Earth Day took root on April 22, 1970, and has since spread across the country as an annual event in thousands of schools, churches, and local communities as well as in many countries around the world.

What was the purpose of Earth Day? How did it start? These are the questions I am most frequently asked. Having spoken on environmental issues in some two dozen states during the twelve years before that first Earth Day, I knew the public was far ahead of the political establishment in its concern for what was happening to the environment. The signs of degradation were everywhere—polluted rivers, lakes, ocean beaches, and air.

The goal of Earth Day was to inspire a public demonstration so big it would shake the political establishment out of its lethargy and force the environmental issue onto the national political agenda. That is what happened.

The idea for Earth Day evolved over a period of seven years starting in 1962. It had been troubling me for several years that the state

Scenes of riverfront waste disposal were the rule before the first Earth Day. *Photograph courtesy of National Archives and Records Administration, from U.S. Environmental Protection Agency.*

of our environment was simply a non-issue in the politics of the country. Finally, in November 1962, an idea occurred to me that was, I thought, a virtual cinch to put the environment into the political limelight once and for all. The idea was to persuade President John F. Kennedy to give visibility to the issue by going on a national conservation tour. I flew to Washington to discuss the proposal with Attorney General Robert Kennedy, who liked the idea. So did the president.

By coincidence, the Senate scheduled a vote on ratification of the Nuclear Test Ban Treaty for the same date—September 24, 1963—that the president chose to begin his five-day, eleven-state conservation tour. The president delayed his departure until I and Senators Hubert Humphrey, Gene McCarthy, and Joe Clark had voted on the ratification so that we could join him for the first leg of the conservation tour. When we took off on *Air Force One,* the plane was loaded with press and TV reporters. But the hot news was ratification of the Test Ban Treaty, which President Kennedy strongly supported. At every stop during the next five days the test ban was what

The Clean Water Act and amendments in later years curbed pollution like that found on this Oregon beach in 1972. Coos Bay District Attorney Bob Brasch examined sludge from a pulp mill settling pond that spilled foul-smelling liquid into the ocean. *Photograph courtesy of National Archives and Records Administration, from U.S. Environmental Protection Agency.*

the news media wanted to hear about—they didn't know much of anything about environmental issues, and their editors knew even less.

Although the tour did not succeed in putting the issue on the national political agenda, it would be the germ of the idea that ultimately flowered into Earth Day. During the next few years, I spoke to audiences all across the country. The evidence of environmental deterioration was all around us, and everyone noticed except the political establishment. The environmental issue simply was not to be found on the nation's political agenda. The people were concerned, but the politicians were not.

After President Kennedy's tour, I still hoped for some idea that would thrust the environment into the political mainstream. Six years would pass before the idea that became Earth Day occurred to me while on a conservation speaking tour out West in the summer of 1969. At the time, anti–Vietnam War demonstrations, called "teach-ins," had spread to college campuses all across the nation. Why not organize a huge, grass-roots protest about what was happening to our environment?

It was a time when people could see, smell, and taste pollution. The air above major cities such as New York and Los Angeles was orange, Lake Erie was proclaimed dead, and backyard birds were dying from a chemical known as DDT. Public interest was further piqued by two environmental catastrophes that captured headlines from coast to coast earlier that year. The first was a large oil tanker spill offshore Santa Barbara that left the public with images of sea birds coated in oil. Then in June of 1969, the Cuyahoga River—slick with oil and grease and littered with debris—caught fire and shot flames high into the air in Cleveland. That image, widely circulated in the popular press, burned its way into the nation's collective memory as the poster child for the environmental atrocities of the time.

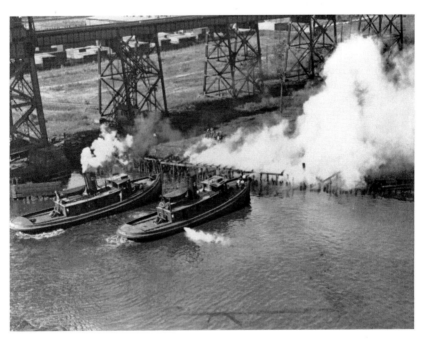

Before Earth Day, river fires were considered the price of progress. Flammable liquids and materials were dumped because of poor regulation and poor enforcement of weak laws. Fires occurred on the Cuyahoga River at least once a decade from the 1930s to the 1960s. An infamous fire broke out on the Cuyahoga River in June 1969, where flames shot high into the air. The incident underscored the nation's environmental troubles and galvanized the public in the months leading up to Earth Day. *Photograph courtesy of Cleveland Public Library.*

For decades, garbage from Manhattan and other boroughs of New York City was barged to Staten Island. The trash was dumped in a major wetlands, transforming it into the nation's largest landfill. After Earth Day and the rise of modern environmentalism, recycling became a national campaign, eventually resulting in federal laws and grants to promote recycling and stop the filling of coastal wetlands. *Photograph courtesy of National Archives and Records Administration, from U.S. Environmental Protection Agency.*

I was satisfied that if we could tap into the environmental concerns of the general public and infuse the student anti-war energy into the environmental cause, we could generate a demonstration that would force this issue onto the national political agenda. It was a big gamble, but worth a try.

At a September 1969 conference in Seattle, I announced that in the spring of 1970 there would be a nationwide grassroots demonstration on behalf of the environment, and I invited everyone to participate. The date April 22 was chosen because it was before the summer recess for grade and high schools, and it avoided exam time on college campuses. I believed the support of these groups would be critical to any successful demonstration on behalf of the environment. That turned out to be a good guess.

The date aroused the suspicions of the conservative John Birch Society, however, which perceived some sinister communist plot was under way. Within a week of the announcement that April 22 would be Earth Day, the society charged that the event was "Sen. Nelson's ill-concealed attempt to honor the 100th anniversary of the birth of Lenin." Obviously, the John Birch Society was better informed about Lenin than I was. This coincidence of timing continued to pop up here and there. The day before the first Earth Day the Los Angeles City Council adopted an Earth Day resolution by one vote over the objection of a member who argued against passing such a resolution on Lenin's birthday. In 1990, while on a speaking tour celebrating the twentieth anniversary of Earth Day, Notre Dame students told me the school had received a letter from the John Birch Society demanding to know why a good Catholic university such as Notre Dame would celebrate the birth of Lenin. If nothing else, the Birch Society should be awarded a medal of distinction for dogged and obtuse persistence.

The wire services carried the story about the planned Earth Day demonstration from coast to coast. The response was electric; it took off like gangbusters. Telegrams, letters, and telephone inquiries poured in from all across the country. The American people finally had a forum for expressing their concern about what was happening to the land, rivers, lakes, and air—and they did so with tremendous exuberance. For the next four months, two members of my Senate staff, Linda Billings and John Heritage, managed Earth Day affairs out of my Senate office.

One of my most immediate problems was to raise money to cover expenses. Within a few days, environmental lawyer Larry Rockefeller came to my office to inquire about the planned demonstration. After he left, one of my staff handed me an envelope from him. It contained a check for $1,000—the first Earth Day contribution. Shortly thereafter, I called two old friends in the labor movement, Walter Reuther, president of the United Auto Workers, and George Meany, president of the American Federation of Labor. They each contributed $2,000. That gave us start-up money. Speaking fees produced another $18,000, and Sydney Howe of the Conservation Foundation came to our rescue with $20,000. More contributions came in from individuals who were supportive of the cause, and we ran the whole show with just $185,000.

Five months before Earth Day, on Sunday, November 30, 1969, the *New York Times* reported that students from Maine to Hawaii were campaigning on environmental ills ranging from water pollution to global population and were rolling up their sleeves. *Times* reporter Gladwin Hill captured the mood of the time:

Rising concern about the environmental crisis is sweeping the nation's campuses with an intensity that may be on its way to eclipsing student discontent over the war in Vietnam. . . . Already students are looking forward to the first "D-Day" of the movement, next April 22—when a nationwide environmental "teach-in," being coordinated from the office of Senator Gaylord Nelson, is planned.

It was clear we were headed for a spectacular success on Earth Day. Grassroots activities had ballooned beyond the capacity of my Senate office staff to keep up with the telephone calls, paperwork, and inquiries. In January, three months before Earth Day, John Gardner, founder of Common Cause, provided temporary space for a Washington, D.C., headquarters. I staffed the office with college students.

The environment was seen as an issue that bridged generational, political, and social gaps. Politicians soon heeded the call. In the months preceding the event, my office received dozens of requests for speeches from congressional colleagues who had never spoken on the environment. When Earth Day finally arrived, Congress adjourned for the day because so many members were participating in the day's events. Again, the *New York Times* spoke to Earth Day's ability to transcend long-established boundaries, reporting that "Conservatives were for it. Liberals were for it. Democrats, Republicans and independents were for it. So were the ins, the outs, the Executive and Legislative branches of government."

Earth Day worked because of the spontaneous response at the grassroots level. We had neither the time nor the resources to organize the 20 million demonstrators who participated from thousands of schools and local communities. That was the remarkable thing about Earth Day. It organized itself.

Every year around April 22, the press abounds with news stories and columns pontificating about how corporations have co-opted Earth Day and are perverting the occasion into a self-promotion about their concern for the environment. One can pull up Earth Day articles from five, ten, or twenty years ago and find the same charges.

They all miss the point of what is actually happening in the real world of changing attitudes, expanding concerns, and public understanding of the environmental challenges we face.

When I planned the first Earth Day, only one major corporation contributed to our national effort—Arm and Hammer. Corporate leadership there was far ahead of the times in its concern about our environmental future. Things have not stood still since 1970. Corporate leadership at the top and middle management levels has gone through the first Earth Day and many since. Certainly not all have become environmentalists. Most are more sensitive to the issue, however, and many others are better informed and committed to forging a sustainable society—defined as a society that meets our present needs without compromising the ability of future generations to meet theirs. It is true that business leadership has a long way to go, but it is also true that it has come a long way since the first Earth Day.

The following article, published by *E Magazine* on Earth Day 1993, had this to say about Earth Day's legacy:

Brace yourself. Earth Day's back and, some would say better than ever. Once again, people will plant thousands of trees . . . thousands will gather for poetry readings, speeches, protests, concerts, recycling contests, more speeches.

And the criticism will flow. Local TV and print media will again dismiss Earth Day as "vaguely reminiscent of the 60s," pointedly noting that events again failed to attract the numbers of 1990. Professional environmentalists parading through shows as disparate as "Donahue," "Crossfire," and National Public Radio's "Talk of the Nation" will undoubtedly remind us that what really matters is what happens the other 364 days of the year. Greenpeace will protest the "greenwashing" ad campaigns of companies like DuPont and GM, two corporations that have aggressively capitalized on the presence of Earth Day. Big business, they will say, co-opted Earth Day.

Both media and many environmentalists will again miss the real point—and the real audience—of Earth Day. TV news action cameras will scour the big events in Washington, San Francisco and New York but will miss millions of smaller ones happening everywhere. Here is the real Earth Day, for April 22 has found a permanent home on the calendar of thousands of schools worldwide. Every significant Earth Day festival features activities geared for kids and in many families it's the kids who bring their parents to the Earth Day celebrations. Big business did not co-opt Earth Day. Kids did.

Earth Day is no longer a special event. It's just part of the curriculum. In many of our nation's schools, Earth Day provides a huge boost to the small but fierce community of environmental educators. It gives visibility to the school, offers teachers the chance for unforgettable teaching moments and lets kids connect their work to their community. On Earth Day, the classroom plugs into the world.

We will not succeed in forging an economically and environmentally sustainable society until all key social, political, economic, and religious groups are on board. If labor, business, or any other major group is opposed to doing what is necessary to achieve sustainability, it probably will not happen.

By the same token, we should not be blind to the fact that, for some companies, Earth Day is merely an occasion to put on their Sunday best one day a year. It is evolving, however. Whether a corporation wants to appear green for public recognition, or for perfectly honest reasons, it doesn't matter. The fact is, we're gaining. If they are deceiving the public, let them be exposed in the political marketplace. Keep them honest if you can. They can read the science as well as we, but an environmentalist can't go out and tell businesspeople that they must mend their ways and expect them to listen.

No one is pure. We all make compromises in different aspects of our lives where the environment is concerned, and it is self-defeating to draw rigid lines of purity when trying to build a political consensus.

In short, the company that buys into Earth Day once a year but that fails to clean up its act plays the same game as the individual who picks up litter one day a year but fails to be a good steward of the planet every other day. Earth Day is not a day of penance for America. It was founded on a spirit of desire and a sense of duty— as a means to an end, not as an end. Let us keep that spirit alive and our goal clear.

Contrary to what some Earth Day critics today might say, my thinking was not that a one-day demonstration would convince people of the need to protect the environment. I envisioned a continuing national drive to clean up our environment and set new priorities for a livable America. Earth Day was to be the catalyst.

The public spoke with one voice at that first event, and its message was heard. The same year, President Richard Nixon created the

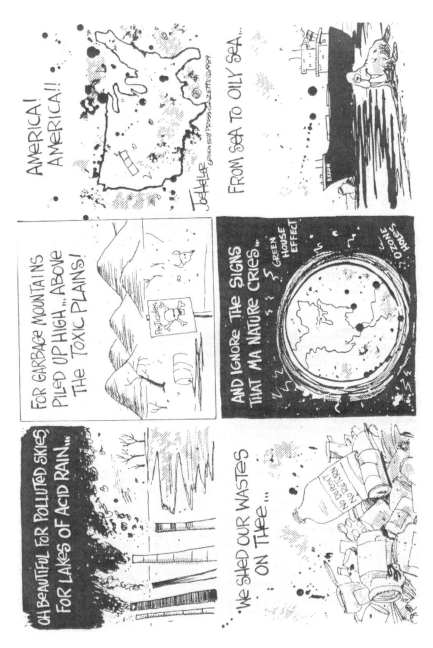

Joe Heller/*Green Bay Press-Gazette*.

U.S. Environmental Protection Agency (EPA), and Congress passed an amended federal Clean Air Act. In the decade that followed, twenty-eight other significant environmental laws were enacted, more environmental legislation than Congress had passed in all the years since the adoption of the Constitution. Some of those laws built upon and strengthened earlier measures, such as the Clean Air Act and the Clean Water Act. Others set the foundation for environmental education in the schools and basic environmental protections that many Americans now take for granted, among them the Safe Drinking Water Act, Endangered Species Act, and Marine Mammal Protection Act (see sidebar).

Major Federal Environmental Initiatives since Earth Day 1970

1970	Environmental Protection Agency is created by executive order
1970	Clean Air Act (1967 act amended)
1971	Alaska Native Claims Settlement Act
1972	Clean Water Act (Federal Water Pollution Control Act Amendments)
1972	Coastal Zone Management Act
1972	Marine Mammal Protection Act
1972	Marine Protection, Research and Sanctuaries Act
1973	Endangered Species Act
1974	Energy Supply and Environmental Coordination Act
1974	Forest and Rangeland Renewable Resources Planning Act
1974	Safe Drinking Water Act
1974	Deepwater Port Act
1975	Eastern Wilderness Act
1975	National Environmental Policy Act Amendments
1976	National Forest Management Act
1976	Federal Land Policy and Management Act
1976	Resource Conservation and Recovery Act
1976	Toxic Substances Control Act
1976	Federal Coal Leasing Act Amendments

1977   Clean Water Act Amendments
1977   Clean Air Act Amendments
1977   Surface Mining Control and Reclamation Act
1977   Soil and Water Resources Conservation Act
1978   Endangered American Wilderness Act
1978   Outer Continental Shelf Lands Act Amendments
1978   Omnibus Parks Act
1979   Archaeological Resources Protection Act
1980   Alaska National Interest Lands Conservation Act
1980   Comprehensive Environmental Response,
       Compensation and Liability Act (Superfund)
1985   Superfund Amendments
1985   Safe Drinking Water Act Amendments
1987   Clean Water Act Amendments
1988   Endangered Species Act reauthorization.

In addition to these major initiatives, hundreds of individual public lands bills have been enacted since 1970 to expand existing national parks, refuges, and forests; to create wilderness areas and wild and scenic rivers; to designate new units of the national park and wildlife refuge systems; and to protect and conserve other important natural areas on the public lands.

That first Earth Day showed that environmental activism on a broad scale was not only possible but powerful, as people across the spectrum of American life demanded that the right to a decent environment be adopted as a fundamental aim of society. The demonstration marked more than a national holiday for the Earth. It was about the people sending a message and setting an agenda. It was about sparking a landmark grassroots movement.

In October 1993, more than twenty years later, *American Heritage* magazine would write this about the event:

On April 22, 1970, Earth Day was held, one of the most remarkable happenings in the history of democracy . . . 20 million people demonstrated their support . . . American politics and public policy would never be the same again.

2

Report Card on the Earth

A casual look at the deterioration that has come about over the past 30 years is a frightening prologue to a disaster of inestimable dimensions if the accelerating rate of the environmental crisis continues. It is not, however, a trend that cannot be reversed.

—1970

Substantial strides have been made in the United States since that first Earth Day in 1970, when the public signaled loud and clear its disgust with dirty air and water.

In the wake of sweeping environmental laws, most notably the Clean Air Act and Clean Water Act, much of the visible pollution has been cleared away. Black smoke no longer billows and curls from smokestacks; raw sewage no longer runs into waterways in the amounts it once did; Ohio's Cuyahoga River no longer burns from pollution; and Lake Erie's fisheries have been revived from a death-like slumber.

On other fronts, prairie restoration is taking place in areas of the Midwest. Coastal protection is under way in California. Waterfowl habitat has been protected and expanded in Louisiana, Alabama, and elsewhere along the Mississippi Flyway. Electricity from renewable sources is now an option for more than one third of all U.S. consumers, and backyard wildlife programs are flourishing in communities across the country.

Yet, as millions gathered in as many as 184 countries to celebrate Earth Day's thirtieth anniversary in 2000, they warned of a world challenged by new, more complicated problems. Here in the United States, smog alerts continue to keep children and the elderly indoors in Atlanta, Houston, and Chicago. Inundating floods still roll along the Missouri and mighty Mississippi Rivers in the absence of flood-plain management. Fish consumption warnings have increased in all of the Great Lakes after years of chemical contamination and have spread to places such as Oregon's Willamette River. Majestic old-growth forests, including the giant redwoods, are disappearing in northern California. Bird populations are waning in the Florida Ever-glades, and all-terrain vehicles are destroying wildlife habitat in our national parks and forests and in the desert lands of the West.

We share one, interconnected planet; as the United States fares, so fares the world in many respects. Our Earth today is shaken each year by increasing assaults from severe storms, heat waves, and floods as the average global temperature rises at an accelerated rate. More and more plant and animal species become extinct with each passing year. Use of the Earth's polluting—and finite—fossil fuels is growing, absent earnest development of alternative renewable sources to replace them. In places such as North Africa and Central Asia, freshwater supplies are shrinking. The rainforests are under siege in Malaysia and Indonesia. Inuit women living in remote regions of the Arctic have high levels of harmful chemicals in their breast milk, and chemicals are also found in the region's polar bears, seals, and fish. Amphibian populations are rapidly decreasing in Australia, Puerto Rico, Costa Rica, and elsewhere, causing scientists to wonder whether humans are seeing their own future foreshadowed in these environmentally sensitive creatures.

These early warning signs are just the tip of the iceberg—and, what's more, scientists tell us that the Earth's icebergs are melting. All are evidence that, some thirty-plus years after the first Earth Day, humankind is exceeding the carrying capacity of the Earth, and the planet is straining under the pressure.

But we don't have to travel the world, or even the United States, to see the toll. On a smaller scale and closer to home the signs are visible all around, if we take the time to look. A hazy ceiling over-hangs our big cities much of the time, blotting out by day the dra-mas of roving clouds and sky and by night the starry heavens.

In smaller cities and towns, the homes and strip malls that once ringed downtowns creep onto rural farmland, woodlands, and wet-lands, carving up the landscape and lacing it with more roads, more traffic. Rivers and streams have been transformed into floods of dirty water in areas where development has bared the Earth, and an intricate, interwoven tapestry of native plants, grasses, and trees has been replaced by imported Kentucky blue grass and hothouse flowers.

Some take heart at the rejuvenation of populations of the once-endangered bald eagle in North America's lower forty-eight and of populations of tigers and pandas elsewhere in the world. Yet in urban backyards across the United States, our wildlife has been disrupted without many of us even noticing. Colorful songbirds are outnumbered or replaced by sparrows, crows, and pigeons, and for many city dwellers, the gray squirrel, raccoon, and occasional rat are the most common wildlife sightings. All are species that thrive and congregate around people, having driven out a richly diverse wildlife community that dies off as humankind now claims the habitat as its own.

But any ecologist will tell you that the growing human population is the greatest single stressor endangering the planet's carrying capacity, and that most environmental ills today stem from overpopulation. The number of people on the Earth tripled from 1930 to 2000, from two billion to six billion. That equates to adding four nations the size of China in a span of sixty years. In the next half-century, we will add the equivalent of three more Chinas, swelling the human ranks to nearly nine billion. The greatest population gains will occur in developing countries such as India, Ethiopia, and Pakistan. Despite a nearly stable replacement birth rate today, the United States could see its population of more than 280 million double in the next seventy to seventy-five years as it makes room for some of the world's expanding numbers. Once we have exceeded the Earth's carrying capacity—and we have when our numbers and consumption patterns result in the erosion of soil, the disappearance of species, and the sullying of our air and water—the planet's health descends on a spiraling downward course.

We can't reverse the destruction, but we can slow and conceivably halt it. It won't be easy. All signs point to the need for greater public understanding of environmental problems and pressures, as we leave behind a century that has done more to damage the Earth than any preceding it. Our legacy over the next one hundred years could be worse. The decisions made even in the next thirty years will guide the course of planetary history like none before—affecting our social as well as our ecological well-being.

Sadly, this is where we are today.

I say this at the risk of being characterized as a bearer of gloom and doom. I've heard it before. It is a tried-and-true gimmick for shutting up or putting down those who speak on behalf of nature. We are a nation of optimists where everything must always be upward and onward. If you're saying *it ain't,* well, you're on the wrong track—you're spoiling the party. This, most definitely, is not the stuff of dinner-party pleasantries. It isn't pessimism, but realism. If we take a realistic look and see what can be done to avoid or alleviate the most serious consequences of our actions with respect to the environment, that's not gloom and doom. The real naysayers are those who refuse to acknowledge that there are many positive things we can do to achieve sustainability. The real naysayers are those who tell us everything is rosy, when hard science tells us it's not.

When environmentalists aren't criticized for being dour, they're accused of having little concern for the economy and "progress." Sometimes explaining things in economic terms helps. I have a friend whose guiding theology for all political matters is the editorial page of the *Wall Street Journal.* He never could quite understand the direct and beneficial connection between a healthy environment and a prosperous economy until I described it in the jargon of his business world. "Look at it this way," I said. "The economy is a wholly owned subsidiary of the environment. All economic activity is dependent upon that environment and its underlying resource base of forests, water, air, soil, and minerals. When the environment is finally forced to file for bankruptcy because its resource base has been polluted, degraded, dissipated, and irretrievably compromised, the economy goes into bankruptcy with it. The economy is, after all, just a subset within the ecological system."

Yet today, as we step with both feet into the new millennium, I see a world in which many of us are becoming more and more isolated from nature and its finite reality. Human kinship with the natural world is being supplanted by interaction with an increasingly developed world, as suburbs sprawl across pastoral countrysides, and more than two dozen major cities follow the likes of New York and Tokyo into the age of the megacity. Three out of every four Americans live in urban environments these days, up from about half the population just before the bombing of Pearl Harbor in 1941.

At the same time it is easy, indeed comforting, for us to embrace the positive messages spun by industry groups and others who profit in the short term from the exploitation of the Earth's resources. I add to this another sobering thought: that the nation's relatively new awakening to environmental stewardship, embodied in that first Earth Day, may have bred a complacency among the public that environmentalists alone would watch over the planet. Indeed, for many today, care of the Earth is best done at arm's length—one day a year.

In retrospect, Earth Day may have had its price. We are losing the battle. You can talk all you please about protecting species diversity, but we're losing it. Slowly. Year by year, day by day. Even as we become better informed and more concerned about the issues, we're losing ground. We can't double the population without doubling the problems. We're becoming disaster managers.

Now, at the beginning of the twenty-first century, Earth Day is reaching its prime, and I, at eighty-six, am considered a senior citizen. But this is no time to rest—for either of us.

PART 2
Imperiled Planet

3

Windows on the World

There is not merely irritation now with the environmental problems of daily life—there is a growing fear that what the scientists have been saying is all too true, that man is on the way to defining the terms of his own extinction.

—1970

It is the challenge and responsibility of every nation to conscientiously seek sustainability to the extent feasible within its own borders. The evidence to date is unmistakable that the environmental damage is serious and mounts day by day. The deterioration is taking shape around the globe—appearing in the form of species extinctions, global climate change, the collapse of ocean fisheries, and more.

All of this is happening here and now. The gravity of these threats is such that within the last decade, the United Nations has held several international conferences to address these most critical problems of our time. The first, the 1992 Earth Summit, saw some 47,000 people around the globe converge on Rio de Janeiro seeking a way to better balance environmental protection and economic development. The Rio summit delivered a powerful message to the world when the participating nations proclaimed: "Humanity stands at a defining moment in history. We are confronted with a perpetuation of disparities between and within nations, a worsening of poverty, hunger, ill health and illiteracy, and the continuing deterioration of the ecosystems on which we depend for our well-being."

But the message was not without hope. The summit underscored the importance of working toward a sustainable world in which concern for the environment and development are adequately weighed and

23

balanced. International conferences—in Cairo, Egypt, in 1994 and Kyoto, Japan, in 1997—built on the Rio summit by focusing on the problems of world population growth and global climate change. Sustainable living again was the prescribed tool, an alternative to our current system of measuring progress on the basis of economic growth.

We in the United States can meet this challenge, given effective political leadership and the will to succeed. Clearly, no nation can achieve this goal entirely on its own, but we can go a long way toward achieving sustainability within our own borders. We cannot wait; we must move now to address those questions we have the knowledge to address.

We are fortunate to live in an era of unprecedented wealth. But we should be mindful that this prosperity comes at a price. The goods we buy, the products we sell, all have their roots in nature's stores. Wood, soil, water, minerals, plants, and fossil fuels are the building blocks of our economy, yet these resources are not of our design.

Humans have been exploiting Earth's resources for as long as they have walked the planet. But this exploitation has been accelerating in the last one hundred years at a tremendous pace. It has been only in relatively recent times that we've learned to dam large rivers to create power and burn fossil fuels to make electricity; to create plastics from petroleum; to fission atoms to make nuclear power. Put these innovations against the vastness of Earth's geologic time, reaching back billions of years, and one begins to sense the limited experience we have with the side effects of our progress.

For many, today's global problems seem fractured, unrelated, and unavoidable. They aren't. Human population growth and activity are at the heart of most of the problems, and we have both the ability and responsibility to address them.

If those for whom environmental issues remain something of a mystery could see the world's ecological ills broadly outlined in one place, with the relationships drawn like a connect-the-dots puzzle, they would begin to comprehend the whole of the problem. Without this understanding, the public cannot fully fathom the magnitude of what is at work, or at stake. Without it, people are vulnerable to false claims of global health by those who stand to profit from its resources. Without it, people find refuge in the possibility that hidden in those murky aspects of environmental issues they don't quite grasp lie answers that would put the problems into proper perspective. Ignorance, in this case, is indeed bliss.

I don't have the expertise to explain the world's environmental troubles. But I can point to the growing body of evidence gathered thus far by experts. Let us part the curtains on our world, if you will, and take a look at what these experts have found.

Population Peril

Population growth is a world peril. By almost any measure, the population of the world exceeds the carrying capacity of the planet. That is to say, we are consuming the world's stores of natural capital to maintain our current standard of living. The resource base that sustains our economy is rapidly dwindling. We are talking about deforestation, aquifer depletion, air and water pollution, fishery depletions, urbanization, soil erosion, and more.

It is worth noting that at this point in history no nation has managed, by accident or design, to evolve into a sustainable society. Clearly, forging and maintaining such a society is the critical challenge for this and all generations to come. In responding to that challenge, addressing rapid population growth promises to be a key factor in determining our success or failure.

When experts are asked to list the most critical environmental problems, they are nearly unanimous in ranking exponential population growth at the top of the list. In a dramatic and sobering joint statement issued in 1992, the U.S. National Academy of Sciences and the Royal Society of London, two of the world's leading scientific bodies, joined for the first time to issue a warning to the world's citizens. They used the following words in addressing the state of the planet:

If current predictions of population growth prove accurate and patterns of human activity on the planet remain unchanged, science and technology may not be able to prevent either irreversible degradation of the environment or continued poverty for much of the world. . . . The future of our planet is in the balance. Sustainable development can be achieved, but only if irreversible degradation of the environment can be halted in time. The next 30 years may be crucial.

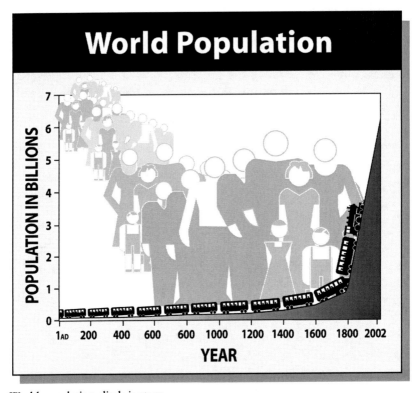

World population climb is steep.
World population growth exploded shortly after 1800 when the world's population was about 1 billion. In 2000 the population passed 6 billion, and another 3 billion people are expected to join us by 2050. For every billion the Earth sustains, an enormous amount of natural resources is used, and conflicts as well as ecological damage are unavoidable. *Illustration*: Chuck Lacasse of Resurgent Creative, Green Bay.

In 2001, the "Amsterdam Declaration on Global Change," drafted by the scientific communities of four international global change research programs, issued an equally stern warning:

Anthropogenic changes to Earth's land surface, oceans, coasts and atmosphere and to biological diversity, the water cycle and biogeochemical cycles are clearly identifiable beyond natural variability. They are equal to some of the great forces of nature in their extent and impact. Many are accelerating. Global change is real and is happening *now*. . . .

The accelerating human transformation of the Earth's environment is not sustainable. Therefore, the business-as-usual way of dealing with the Earth System is not an option. It has to be replaced—as soon as possible—by deliberate strategies of good management that sustain the Earth's environment while meeting social and economic development objectives.[1]

Is there any other issue even a fraction as important? Yet, the leaders of this country remain silent about sustainability and the disastrous consequences of exponential growth.

There are more people alive today on the Earth than have died throughout all of human history. How did this happen?

In his book, *Billions and Billions,* the late astronomer Carl Sagan explained the power of exponential growth as it relates to human birthrates, noting that today the world population doubles about every forty years. This explosive population growth follows a relatively steady state in which the number of births and deaths were nearly balanced

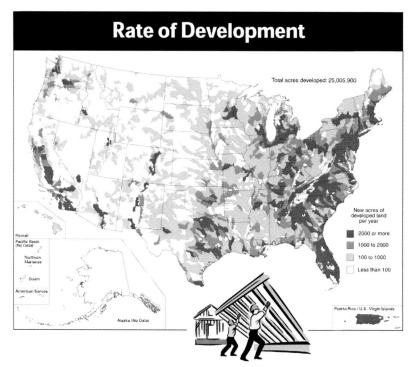

Rate of development hits southeast hardest.
People need land to build homes and run businesses, but where the land is and how much is used is a choice, not a need. The rate at which land is converted is rapid, and has greatly accelerated every decade since Earth Day. In the 1980s and 1990s the greatest rate of conversion occurred in the Southeastern United States.

throughout most of the hundreds of thousands of years of human history. The change came about following the advent of agriculture about 10,000 years ago and with it the ability to plant and harvest grains to feed people in numbers not seen before.[2]

The first agricultural revolution transformed the planet, leading to steep increases in the number of people. Scientific and medical advances have furthered that growth, lowering infant death rates at one end of the human life spectrum while raising life expectancy at the other end.

No period of human population growth has been as striking as that of the last century. It took all of human history for the world to reach a population of 1.6 billion at the beginning of the twentieth century. Just one hundred years later, at the century's close, that number nearly quadrupled to more than 6 billion. The addition of the last billion shows the power of exponential growth as it was accomplished in just a dozen years—between 1987 and 1999—and faster than at any time in human history.

The United Nations estimated that the world population reached 6 billion on October 12, 1999. The day arrived amid media hoopla in which experts warned of devastating environmental dangers as the global population continues to rise. At the same time, however, many trumpeted the fact that for the first time, birth rates—the number of children born per woman—had slowed in every nation of the world.

The slowdown is attributed to a variety of factors. Chief among them are that education and greater job opportunities for women, access to family planning, and the growing trend toward urbanization have led to delayed childbirth and smaller families.

This is good news. Yet the decline in birth rates worldwide doesn't diminish the fact that the world is far from stabilizing its population. Indeed, the United Nations' mid-range projection is that world population will continue to mount until it levels off at nearly 9 billion in 2050. Its high-range projection has population approaching 11 billion during that same period. A more optimistic UN scenario is that the number of people on the planet could start to decline in 2040 after peaking at 7.5 billion. This scenario is considered unlikely, however, as it assumes that couples around the world will move quickly toward replacement level fertility, in which they bear only two surviving children.[3] The United Nations has gauged population growth since 1947 and has a credible track record to date, having accurately predicted back in 1970 the timing of when the world would hit the 6 billion mark.[4]

Why the continued growth? Although populations have stabilized and even declined in many Western, industrialized countries, some developing nations are expected to double or triple in size during the next half-century. Ethiopia, for example, is expected to more than triple in population, from 64 million in 2000 to 187 million in 2050.[5] During the same time period, India—which in the last half century saw its population triple to one billion—will grow by another half, adding a projected 515 million people and overtaking China as the world's most populous country.[6]

And demographers say growth in the twenty-first century will differ significantly from that of the last century. Nearly all the growth will be in the urban areas of Third World countries—poor nations with already stressed environments and economies—whereas twentieth-century growth occurred in both the industrialized and developing countries. Already, the UN estimates that 2.6 billion of the world's people lack basic sanitation and 1.3 billion lack access to clean water. These sad statistics can only grow in the face of continued population growth.

Clearly, more people mean more pressures on Earth and its resources, and on our own quality of life. Already, our numbers are upsetting and harming the global atmosphere and climate, the survival of other species, and the availability of resources such as forests, range land, and clean air and water. The UN reported at the 1999 Hague Forum that humans have transformed between one-third and one-half of the planet's entire land surface. As part of that equation, the UN reports that more than a fourth of the world's birds have been lost, and about three-quarters of the major marine fisheries have been fully exploited or depleted.[7]

We've talked about the exploding world population. What is not yet generally recognized, however, is that the United States population is the fastest growing of any industrialized country and is unmatched in its consumption of resources.

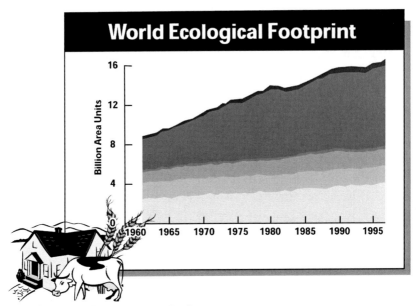

Human ecological impact is growing larger every year.
The worldwide ecological footprint—a measure of humanity's ecological pressure on the Earth in terms of our consumption of food, materials, and energy, and the corresponding waste—has nearly doubled in the last forty years. Over the long term, this rate of depletion far exceeds the biosphere's ability to sustain it. *Illustration:* Chuck Lacasse of Resurgent Creative, Green Bay.
Source: World Wildlife Fund for Nature, *Living Planet 2000* (Gland, Switzerland: World Wildlife Fund for Nature, 2000); Nicky Chambers, Craig Simmons, and Mathis Wackernagel, *Sharing Nature's Interest: Ecological Footprints as an Indicator of Sustainability* (Sterling, Va.: Earthscan, 2000).

At the current rate, our population will approximately double to more than a half-billion people within the next seventy to seventy-five years. That growth will bring about tremendous changes to this country within the lifetime of our grandchildren, yet the trend of U.S. population growth continues to be largely ignored.

It has been studied, however. A little-publicized 1996 report from a task force of the President's Council on Sustainable Development concluded that population stabilization is one of the most important goals this country could achieve. Formed to develop recommendations that would help move the United States toward achieving sustainability, the council's Population and Consumption Task Force wrote: "The size of our population and the scale of our consumption are essential determinants of whether or not the United States will be able to achieve sustainability. U.S. population and consumption trends demonstrate that a great deal of work needs to be done."[8]

The United States is the only major industrialized country in the world today where large-scale population growth is occurring. We grow by about 1 percent per year, adding 3 million people annually—the equivalent of another Connecticut each year, or a California each decade. Only a handful of countries in the world, all of them developing, add more people to their populations on an annual basis.[9]

It should be noted that although an annual population growth rate of 1 percent sounds small, it is, indeed, very large. The percentage growth rate divided into 70 will provide the population doubling time—known as the Rule of 70. Thus, a 1 percent growth rate divided into 70 tells us our population will double in seventy years. A 2 percent growth rate equals a doubling time of thirty-five years. Some countries are doubling their population in twenty years or less.

The U.S. birthrate is now at what is known as replacement rate, just a slight fraction more than two children for every woman of child-bearing age. Over a relatively brief time, this would stabilize the U.S. population—except for the current immigration rate. Immigration—legal and illegal—adds significantly to our population, contributing a third of the 3 million new residents inhabiting this country each year.

At the current rate, our population will double to more than 500 million people well before the end of the twenty-first century and will balloon to more than a billion sometime in the next century. What does all this mean in practical terms? It means we'll need to at least double the total infrastructure of the United States, and do it all in the next seventy to seventy-five years. We will need to double almost everything, including the number of airports, highways, grade schools, high schools, colleges, apartment houses, homes, hospitals, prisons, and more. It will mean more sprawl, longer commutes, more megacities of 10 million people or more. It will mean at least double the pressure on every resource we have.

Will the population pressure of a billion people force us to consider the much-criticized Chinese policy of one child? After all, both countries are roughly the same size geographically.

People can say all they want about the out-of-control growth in far-flung, Third World countries; blaming population growth on developing countries has always been an easy way for politicians to shirk responsibility for what is happening in the United States. We should help, of course, with education, family planning, technical aid, economic development, and the like. But each country has to solve its own population problem, and in the end, the only country we have any control over is this one.

We must face up to our own problem here at home. We can't wait. In my own lifetime, we have seen this country's population swell from about 98 million when I was born in 1916 to 132 million when World War II started more than two decades later. Today, it has soared to more than 280 million.

As we crowd our lands, we affect our cities, our landscape—even the fish in the sea. Consider the following:

- *Fisheries:* Internationally, the United Nation's Food and Agriculture Organization reports that up to 50 percent of the world's major fishery resources are fully exploited, and about a quarter are overexploited or depleted.[10] Off the North American coasts at least 82 species or distinct populations are vulnerable, threatened, or endangered. These include sharks, sawfish, skates, sturgeons, Pacific coast smelts, cod, seahorses and pipefishes, rockfishes, snooks, and groupers—even the Atlantic salmon and Atlantic halibut.[11]

- *Urbanization:* Sprawl is a problem wherever land use planning laws are lacking. The New York City area saw its regional population grow 8 percent between 1970 and 1990, while land used for the urban area grew 65 percent. In Portland, Oregon—where a regional planning system is in place—population grew 50 percent from 1975 to 1995, while land used for this population grew only 2 percent.[12]

- *Roads:* Public roads cover 1 percent of the United States, an area the size of South Carolina. One study found that environmental impacts from U.S. roads—traffic noise, salt, and the spread of nonnative plants—reach much farther, affecting about 20 percent of the nation's land.[13]

- *Cars and trucks:* More than 215 million passenger cars and light trucks travel the United States, driving nearly 2.5 trillion miles each year.[14] Vehicle travel in the United States doubles every twenty years.[15] (With demand for motorized vehicles expected to rise as per capita income quadruples in China, the world population of road vehicles will pass 1.1 billion by 2020—or 425 million above the 1996 level, predicts the U.S. Energy Information Administration in its International Energy Outlook 1999. The projected increase will rival the 1996 vehicle population of all the industrialized nations combined.)

It's true that population density in the United States doesn't rival that of Europe: the predicted doubling of our nation's population would bring us to about only one-fourth the current population density of Germany and the United Kingdom.[16] Beyond the fact that our vast mountain ranges and deserts render much of the land in the United States unsuitable for farming and development, we have to ask ourselves, however, whether European levels of crowding and density are the goal here at home.

It bears repeating that anytime we are degrading the environment—eroding the soil, polluting the water, overdrafting the fisheries, overcutting the forests—we are exhausting capital and living beyond our means.

How many people can the United States accommodate and still remain ecologically sustainable? As with all population estimates, such figures involve a fair amount of conjecture, and experts have been trying to get a handle on that number since the 1970s.

Several studies during the 1990s by Cornell University scientists determined that a maximum sustainable population for North America should be around 200 million, and 200 million in South America, for a relatively high standard of living.[17] That conclusion and the one reached in 1996 by the president's task force echo an earlier report issued while I was still in the Senate, not long after the first Earth Day. That report, by the President's Commission on Population Growth and the American Future, recommended that the United States move vigorously to stabilize the population—then at 200 million—as quickly as possible.

In a letter to President Richard Nixon and Congress in 1972, commission chairman John D. Rockefeller III wrote: "After two years of concentrated effort, we have concluded that, in the long run, no sub-

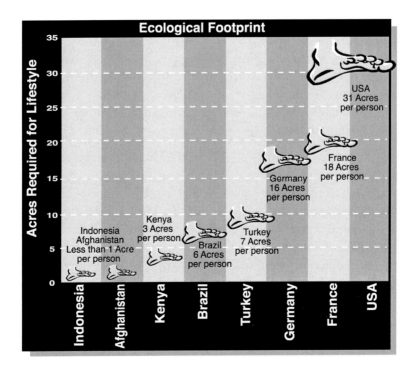

The ecological footprint varies by nation.
The lifestyle we live and the efficiency of our technology determine the ecological pressure—or ecological footprint—we place on natural resources. The average American requires about 30 acres of the Earth to supply resource needs, compared to about 15 for a citizen of industrialized Germany or France, 2 to 3 acres for Brazilians and Kenyans, and less than 1 acre a person for Indonesians and Afghanis. Because of our imports, a large portion of the American footprint is pressure on ecosystems outside the United States.
Source: World Wildlife Fund for Nature, *Living Planet 2000* (Gland, Switzerland: World Wildlife Fund for Nature, 2000); Nicky Chambers, Craig Simmons, and Mathis Wackernagel, *Sharing Nature's Interest: Ecological Footprints as an Indicator of Sustainability* (Sterling, Va.: Earthscan, 2000).

stantial benefits will result from further growth of the nation's population, rather that the gradual stabilization of our population through voluntary means would contribute significantly to the nation's ability to solve its problems."

That report was issued thirty years ago. Since then the population has increased by 80 million in this country, and no public dialogue has occurred on this important issue. It is time to address the issue.

The Sixth Extinction

Ever since 1995, when a group of schoolchildren near Henderson, Minnesota, discovered deformed frogs in a farm pond during a class science project, the subject has grabbed the attention of the national media and the public. "Fully 50% of the frogs caught that day had deformities of their hind legs," according to one account of the students' discovery. "Many had one leg which was underdeveloped and webbed together, preventing normal function. One frog captured that day had only one hind leg, while another had two feet on one leg and a bony protrusion from the spine."[18]

Although frog deformities have been found for centuries, the alarm was sounded when they began appearing in far greater numbers than ever before. As many as fifty-two species of freakish frogs with missing, twisted, or extra limbs, missing eyes, shrunken reproductive tracts, and inside-out organs are now documented in forty-six states and four Canadian provinces.[19]

Scientists are baffled.

At a 1998 Great Lakes conference on frog deformities and amphibian declines, visitors to the Milwaukee Public Museum shuffled by dinosaurs on display, marveling at monsters that roamed the Earth long before humans arrived. But the real monsters were featured inside the museum's darkened auditorium, where a grim freak show played out on a giant screen—grim because this was life today, but not as nature intended. Scientist after scientist approached the podium during the conference, each showing a litany of frog mutations more troubling than the last. A number of potential causes were discussed, from pesticides and other chemicals in the environment to more ultraviolet light from the sun penetrating the thinning ozone layer to predators, parasites, and disappearing habitat as humankind continues its spread. By the end of the day, however, one thing was certain: none of the more than 275 scientists and graduate students assembled had a clear idea what was behind the disturbing trend. "It's depressing," one scientist said in summing up the collective body of knowledge. "The question is: What's happening here? The answer is, we don't know."

One needn't be a fan of frogs—though I certainly am—to pay heed to this problem and what it might portend for other planetary creatures, including people. Scientists worry that in the frogs, we humans may see our own futures foreshadowed. Frogs' permeable skin and vulnerability during several developmental stages make them uniquely sensitive to the environment. Thus, frogs are a sentinel species that act like the old coal miners' canaries, alerting us to danger ahead. Might these frogs, scientists wonder, be an indicator of what's in store for humans and other less sensitive species if today's environmental ills continue?

Years later, answers to the frog question continue to elude scientists. But what is important about the frog deformity story, however, has been its ability—through graphic pictures of misshapen, monster frogs—to attract media and public attention to a well-documented and larger environmental problem: the rapid extinction of thousands of species from our planet.

As Gary Casper, a herpetologist at the Milwaukee Public Museum who coordinated the Great Lakes conference, said, "The fact that we've got frogs coming up with these gross deformities has kind of been a hook, kind of a sensational example of why we need to be concerned. Globally, we are in a crisis with species biodiversity."[20]

Scientists for several decades have been studying dramatic declines and disappearances in amphibian populations around the globe. But frogs and amphibians aren't the only species suffering losses. Plant, bird, and animal life around the world is being obliterated.

In the United States alone, 1,244 species were listed on the nation's endangered species list at the end of 2000.[21] In an attempt to tally global losses, an international consortium of mostly academic and government representatives is compiling a "Red List" of at-risk species—classifying a species as threatened if it has a high probability of extinction during this century. The International Union for the Conservation of Nature reports that 9,477 animal species and 7,022 plant species are known and documented as being at serious risk of extinction—a figure that doesn't include subspecies, varieties, or isolated populations.[22]

Some scientists estimate that extinctions over the last century have accelerated to at least one thousand species per year, up from the natural rate of one to ten.[23] But others say projections of the extinction rate are hampered by the fact that no one really knows how many species currently exist, let alone existed in the past. Estimates of species around the world range anywhere from 5 million to 100 million, with 12.5 million proposed as a reasonable working estimate. Around 1.75 million species are documented as living on Earth today, but scientists tell us we have only begun to discover all the species in large part because the majority of them live in the oceans.[24]

Extinction is a part of life on this planet. Five mass extinctions are known to have occurred during Earth's 4.5-billion-year history, most of them traced to natural phenomena such as meteorite impacts. Scientists say we are now observing the Earth's sixth extinction. There is a key difference this time, however: this sweeping loss of life is caused by humans.

The extent of the threat to the world's biodiversity is considered so great that a binding international treaty designed to protect endangered species was adopted at the 1992 Earth Summit in Rio de Janeiro. Six years later, a survey commissioned by New York's American Museum of Natural History found that among four hundred biologists, 70 percent agreed that animal and plant life is in "the midst of a mass extinction." The same survey found that in stark contrast to the scientists' fears, the public is largely unaware of the loss of species and the threats it poses to human existence in the next century.[25]

The bulk of the world's extinctions will occur outside the United States, wiping out plants and insects in the tropics—home to the planet's richest variety of species. Tragically, thousands of species already have become extinct without our ever having known they existed. Trees, mosses, snails . . . gone with them are potential medicines and other possible benefits they may have had for us, beyond the simple joy of living in a world made more beautiful by its diversity. More important than the potential profit from these species, however, is the silent and mysterious role each plays in the intricate web of life.

Habitat destruction is the leading cause of species die-offs around the globe. The Earth's warming climate, its thinning ozone layer, air and water pollution, and the introduction of non-native species are also said to be major contributors. The spread of people across the landscape and the increasing demands we make on the planet to fuel the growing global economy underlie all of these factors.

The evidence is in: we are doing great damage to many of life's vital ecosystems. This assault on our planet comes from within, not without, and it is all-encompassing. The forests we're clearing, the wetlands we're draining, the chemicals and wastes we're emitting— all are working to damage the Earth at myriad levels from which it cannot easily bounce back once we have pushed it to the brink.

Tallying Our Losses

Here in the United States, we are consuming the habitat of wild creatures at a faster and faster pace. This country has made progress in protecting some of the endangered species that have started along the path to oblivion. Such progress and awareness are, in large part, owing to passage of the federal Endangered Species Act in 1973 and similar state laws that resulted from the environmental consciousness-raising of that time. We have worked to restore populations of the wild bison, the elk, and the timber wolf that at one time or another have been pushed to near-extinction. But progress is slow. For every success story, for every animal, bird, or plant removed from federal and state endangered species lists, more are added.

And there are many less charismatic species battling for their lives in this country, many of them tiny, unattractive, hairless critters that aren't big draws on the television nature shows.

Plants are hard hit, with nearly a third of all plant species in the United States believed to be at risk. In its first large-scale assessment of the nation's ecological wealth, the U.S. Geological Survey reported in 1999 that many of the nation's bird, aquatic, and animal species also are at risk. This assessment found that there are many more species about which we know little that also appear to be in trouble, among them some populations of amphibians and reptiles, invertebrates, small mammals, and ocean dwellers.[26]

A 1999 Canadian study found that North American freshwater species—from fish and frogs to salamanders and snails—are dying out as fast as tropical rainforest species. The study reported that since 1900 at least 123 freshwater species have been recorded as extinct in North America, and warned that the United States could lose most of its freshwater species in the next century if steps aren't taken now to protect them.[27]

The factors behind the plant and animal extinctions in North America are many and, as is true globally, human driven. Intensive land use and pollution, along with the intrusions of non-native species and major alterations to our waterways, are all said to have tremendous impact on fish and wildlife in this country.

The wholesale damming and diverting of waters for irrigation, navigation, and hydroelectric power—with little or no regard to

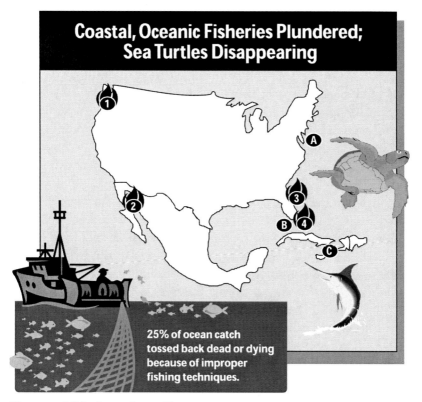

Coastal, Oceanic Fisheries Plundered; Sea Turtles Disappearing

25% of ocean catch tossed back dead or dying because of improper fishing techniques.

The oceans' fisheries are in trouble.
The American Fisheries Society reports that at least four primary "hot spots" exist for U.S. coastal fisheries. Important fish populations and some species face severe depletion or extinction. Hot spots include: 1) Puget Sound, Washington, and adjacent Canadian waters; 2) Gulf of California; 3) Indian River Lagoon area of Florida; and 4) Florida Keys and nearby Florida River. Overfishing, the destruction of wetlands and coral reefs, and destructive new fishing technologies are major causes. Also a problem is the catching of nontarget fish, turtles, and other creatures. All six sea turtle species in U.S. waters are now in small numbers and facing loss of nesting areas. Areas where sea turtles face serious problems include a) Long Island Sound, b) Gulf of Mexico, and c) U.S. Virgin Islands.
Source: U.S. Fish and Wildlife Service; American Fisheries Society

consequences—has altered aquatic systems to such a degree that aquatic organisms are considered imperiled species in this country. In the Southeast and Southwest, for example, 19 percent and 48 percent, respectively, of the two regions' fish species are threatened because waterways have been altered by dams, groundwater pumping, and pollution.[28] Along the Atlantic and Pacific coasts, new super-vacuum technology, sophisticated seines, and electronic spotting techniques are depleting many fish and shellfish populations, prompting calls for consumers to avoid such favorites as lobster, Atlantic cod, and haddock.[29]

As we grow and sprawl to accommodate an expanding population that is expected to double in this century, more land is developed for homes and agriculture, wetlands are drained, and forests are fragmented. Such actions threaten the viability of the same natural life support system that is so critical to people and wildlife alike.

Even our national parks are not immune. Species are disappearing, too, from large protected areas, such as the 1.5 million-acre Everglades National Park in Florida. Biologists estimate that the number of wading birds there has decreased by 90 percent since the 1950s, when the U.S. Army Corps of Engineers built canals and levees to prevent flooding, to drain large sections of the freshwater marsh for farming, and to tap the waters for South Florida's cities.[30] I've been to the Everglades at least fifteen times since first visiting it in 1963, and I can't help but notice the dramatic loss of bird life during that period.

True, much has been done in this nation to protect its biodiversity. But not enough, given the size of the problem, say those who keep the tally sheets. If, as some experts say, the rate of species extinctions is the single best indicator of the health of the planet today, one doesn't have to be a genius to understand the latest global charts. As herpetologist Gary Casper sees it, "The following generations, I think, are going to look back on this period as one of the greatest follies of mankind."[31]

The Earth Is under the Weather

A consensus exists among the majority of the world's climate experts that the Earth's temperature is rising, and that it is rising at a faster rate than it has for a century or more.

The National Research Council—an arm of the U.S. National Academy of Sciences and one of the world's most authoritative scientific review bodies—agrees. The council concluded in early 2000 that "The warming of surface temperature that has taken place during the past twenty years is undoubtedly real, and it is at a rate substantially larger than the average warming during the 20th century." The academy noted further that the acceleration of the Earth's warming in the last two decades has been substantially larger than the average warming during all of the last century.[32]

Indeed, in the last century, the Earth's average surface temperature has climbed an estimated 0.7 to 1.5 degrees Fahrenheit. The U.S. National Climate Data Center, the World Meteorological Organization (WMO), and Britain's national weather service—the Meteorological Office—all report that the 1990s was the warmest decade on record since consistent instrument measurements started back in the mid- to late 1800s. Further, using measurements that predate thermometer readings, such as tree rings and underwater sediment, the WMO reported that the twentieth century was the warmest century during the last one thousand years.

It's worth noting that throughout the 1990s, a bitter debate occurred in the United States about whether the Earth's temperature was on the rise. Meanwhile, concern about the planet's warming climate spawned worldwide conferences addressing global climate change, beginning with the 1992 Earth Summit in Rio de Janeiro. It was there that many of the world's nations approved the goal of reducing so-called greenhouse gas emissions to 1990 levels by the year 2000. A 1997 follow-up conference in Kyoto, Japan, resulted in international agreements to cut fossil fuel emissions by varying amounts after 2000.

Scientists worry that mankind is speeding up the planet's warming. Their concern is rooted in the fact that the level of carbon dioxide (CO_2) in the Earth's atmosphere has risen dramatically since the start of the Industrial Revolution less than two hundred years ago. The United Nations Intergovernmental Panel on Climate Change (IPCC), which is composed of more than 2,500 scientists representing nations around the world and has studied global warming for the last decade, reported in 2001 that the atmosphere's current carbon dioxide concentration has not been exceeded in the past 420,000 years. In fact, the panel reported it likely that carbon dioxide levels haven't been higher at any time during the past 20 million years. Just as important, we are now experiencing a rate of increase that is unprecedented in at least the last 20,000 years of the Earth's history.[33]

Carbon dioxide is known as a "greenhouse gas" because it absorbs energy in the form of infrared radiation emitted from the Earth and thereby warms the Earth's atmosphere. This and several other so-called greenhouse gases, including water vapor and methane, are critical ingredients for making the Earth inhabitable by people and other creatures. As with anything, however, there can be too much of a good thing, and this is the case with carbon dioxide and methane.

Scientists tell us the current atmospheric buildup of carbon dioxide is, in large part, the result of a growing population's increased reliance on burning fossil fuels in the last century. We put carbon dioxide in the air anytime fuel is burned, for everything from heating and cooling our homes to running our cars to powering our industries and businesses.

Today, the world's population is releasing 6 billion tons of carbon dioxide into the atmosphere each year through the burning of fossil fuels. The United States, with just 5 percent of the world's population, is responsible for a quarter of those emissions.[34] Some scientists warn that the Earth's warming will accelerate unless our global emissions of carbon dioxide are greatly reduced—immediately. It is too late to stop the climb in temperature, but we can lower the peak at which the increase will top out. Based upon current trends, scientists say we will emit enough to cause a doubling of carbon dioxide in the atmosphere in the next fifty to seventy years. Already, scientists say the question no longer is, can we stop that doubling, but can we prevent carbon dioxide from tripling in the next one hundred years?

Are We to Blame?

Some question remains about just how much our human activity influences global climate change, an important question because it raises the issue of what we should be doing differently. In a landmark 1995 report, the IPCC caught the world's attention when it concluded, "The balance of evidence suggests there is a discernible human influence on global climate."

Despite critics' attempts to cast doubt on that finding, six years later a working group of the IPCC pointed to even stronger evidence linking the increased buildup of carbon dioxide in the Earth's atmosphere to our warming climate. "In the light of new evidence and taking into account the remaining uncertainties, most of the observed warming over the last 50 years is likely to have been due to the increase in greenhouse gas concentrations," the group wrote in early 2001.[35]

At around the same time in the United States, a National Research Council review of the science of climate change, which was convened to answer questions about global warming posed by the Bush administration, reached a similar conclusion. "Greenhouse

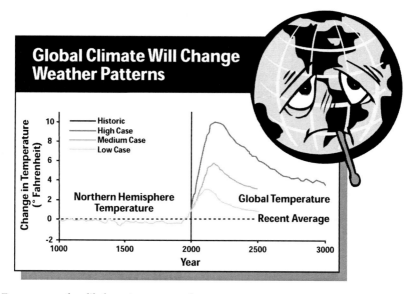

Extreme weather likely to increase as planet warms.
Scientists say human releases of greenhouse gases such as carbon dioxide are tipping the global temperature balance. Our climate is heavily influenced by the Earth's average temperature. The world's top climate researchers are nearly unanimous in warning that the planet's average temperature is rising and will continue to rise rapidly from our air-polluting ways. Scientists predict that hurricanes, floods, and other extreme weather events will occur more often as the Earth's temperature rises. *Illustration*: Chuck Lacasse of Resurgent Creative, Green Bay.

gases are accumulating in Earth's atmosphere as a result of human activities, causing surface air temperatures and subsurface ocean temperatures to rise," the committee wrote. "Temperatures are, in fact, rising. The changes observed over the last several decades are likely mostly due to human activities, but we cannot rule out that some significant part of these changes is also a reflection of natural variability."[36]

What other evidence is there that the climate is warming? Climatologists say the 1990s are noted for featuring the seven warmest years on record, and 1999 was the twenty-first consecutive year recording an above-normal global surface temperature. They tell us further that the Earth's warming climate has manifested itself in a number of ways through the years: Siberia is warmer than at any time since the Middle Ages, Alpine glaciers have retreated, and shrubs are growing on what was previously barren tundra in Alaska.[37]

There are other changes under way as well, everyday differences many of us are likely to have observed in our own backyards. The timing of the birds' arrival in spring, the plants' first blossoms, the spring thaw, and the winter freeze tell us subtly yet certainly the world around us is changing.

Nature's comings and goings have been recorded over the course of sixty years now at the old restored farm near Baraboo, Wisconsin, that was the inspiration for naturalist Aldo Leopold's book *A Sand County Almanac*. From 1935 to 1947 Leopold tracked when the birds arrived and the plants bloomed each spring. His son and daughter, who since 1976 have continued and expanded his recordings at the Leopold Memorial Reserve to track the cycles of some three hundred spring events, report that at least one-third are occurring earlier these days because of the warmer temperatures.

"It's very interesting that a lot of the items my father kept track of in the 1940s are now coming three to four weeks earlier," said Leopold's daughter, Nina Leopold Bradley, whose work with brother Carl was published in 1999 in the *Proceedings of the National Academy of Sciences*. Among their findings is that spring events in the area of the reserve have advanced at a rate of 1.2 days per decade—occurring about a week earlier than in the 1940s. The region's popular sandhill cranes are arriving earlier, for instance, making their first appearance at the farm in mid- to late February compared to the average of March 10 Aldo Leopold recorded back in the 1940s. And migratory Canadian geese making their way north are appearing around February 28, as opposed to around March 22 in the 1930s.[38]

These findings are ominous to Bradley. "It's the greatest threat we have come to yet, especially in terms of biodiversity," she says, recalling a paper she'd read recently about the ecological collapse of the Aleutian Islands. "Because of one species going extinct—the whole food chain went down the drain. Now that's in a very simplistic ecosystem, so I think it's a tremendous threat."

Yet scientists say we have not yet begun to sweat the effects of global warming. World climate models project that if the current trend continues, the global temperature will rise an average of 4.5 degrees Fahrenheit from its current 59 degrees Fahrenheit by the end of the twenty-first century. That change may seem small, but not when one considers that even a slight variation in the average global temperature can cause a dramatic climate shift. A warm-up of 5.4 degrees Fahrenheit or more in the Earth's average temperature would be comparable to the temperature change that occurred between the last major ice age ten thousand years ago and today.[39]

What might happen if the world continues to warm at this pace? Scientists predict any number of life-altering consequences. A chief concern is that global warming could disrupt the atmosphere and oceans that regulate our weather. The United Nations IPCC predicts that global warming could cause a rising incidence of floods, droughts, fires, and heat outbreaks in some regions as temperatures rise. Such changes could threaten water and food supplies in some parts of the world and bring an increasing number of weather disasters—such as hurricanes and severe winter storms—to others.[40]

Rising sea levels also could cause major troubles, directly and indirectly, as water expands in response to warming temperatures, mountain glaciers melt, and the ice sheets melt near the Earth's North and South Poles. Scientists say that even a relatively modest rise in sea level could batter coastal communities around the world. Coastal aquifers would be threatened by invading saltwater, farmland would be lost, portions of some cities would be made uninhabitable, and many populated islands would be inundated. The U.S. Environmental Protection Agency (EPA) estimated that a one-foot rise in sea level along the Atlantic and Gulf Coasts is likely by 2050 and could occur as early as 2025.[41]

Still, some people say they welcome the specter of a warmer climate in their hometown. This is especially true in places like my native Wisconsin—where temperatures can plummet to thirty below and linger there for days. Warmer winters? Longer summers and an extended growing season? Bring 'em on, these folks say.

But warmer temperatures can have less desirable consequences. Scientists say global warming can bring to a region such weather extremes as flooding and drought. That, in turn, can change the local flora and fauna as native plants and animals struggle to keep pace with a rapidly changing environment. Although adapting to the world's continually evolving climate has allowed species to survive through the ages, the rate of warming now under way has scientists worried that species may not adapt or move fast enough.

People, too, have a tougher time adapting to a new climate the faster it changes. Many of us cope by turning up the air conditioning in our homes and offices when a heat wave strikes. But are our cities' infrastructures built to handle the growing demand that will come as more and more people seek relief from warmer temperatures and longer heat waves? The heat wave of 1999, for example, saw parts of Manhattan blacked out when the electric power grid was strained by an unprecedented demand for air conditioning.[42] The upper Midwest had record-breaking power usage in 1998 that caused power to be shut off in parts of that region, and more than five hundred people died during a 1995 Chicago heat wave.

I am not an expert on global climate change. Even the experts, in some respects, are flying blind based on the fact that their map is a climate record that has rich detail for only the past 130 years. Is it possible they could be wrong about the extent to which humans are altering the climate? Yes, it's possible. Science is about figuring things out, and that process involves forming hypotheses and testing whether they are right or wrong.

Perhaps more important, though, is the fact that science is never definitive—never quite final. Witness the theory of evolution. There's always more of what we don't know than what we know, and as we learn, the picture inevitably changes. Most climate experts agree that much more research is needed to understand the intricate workings of our climate system and our influence on it. But that shouldn't scare us off from taking seriously what we do know today and acting on it. Enough evidence is in to convince a majority of the world's climate experts that the Earth is warming faster than it should and that our dependence on fossil fuels has a hand in it. Their warning: we must act now to dramatically reduce our greenhouse gas emissions. If we wait for final, irrefutable evidence, it likely will come too late for us to turn things around.

A strong signal comes to us from within the scientific community itself. Believing that plans to cut fossil fuel emissions to the degree called for in the Kyoto agreement—which the United States walked away from in 2001—won't be enough to slow climate warming, scientists are shifting gears. Finding ways for people to adapt to a warming world is as critical now as mitigating the amount of greenhouse gases pumped into the atmosphere, they say—a telling acknowledgment of how much damage has been done to the Earth's fragile atmosphere already.

What do we do about global climate change? A group of economists, energy company presidents, and policy specialists set out a plan in 1999 to speed up the process called for in Kyoto. The plan was predicated upon the premise that global fossil fuel emissions must be cut about 70 percent—an order of magnitude greater than what was called for in the Kyoto Protocol—in order to allow the planet's climate to restabilize.[43]

Subsidies for fossil fuels in this country and abroad also should be eliminated. This would help curb the short-sighted use of these fuels by giving producers and consumers a more realistic price for gas, coal, and oil. U.S. subsidies for fossil fuels amount to about $21 billion a year, depending on who does the accounting.[44] In this economy, you can buy gas, coal, and oil cheaper than I can compete with photovoltaic cells to harness energy from the sun.

Ultimately we'll have to turn to solar energy and other alternative technologies. If we give the same kind of subsidies to these technologies that we have given to other energy sources, we can make them move. After all, the sun is there. Indirectly and directly, it's the source of nearly all the energy we use today—and we can capture loads of it if we try.

Consider these compelling climate facts compiled by Ross Gelb-span, who has written extensively on the subject of global climate change:

- Number of years that have set heat records since 1983: 10

- Portion of the Kyoto planet protocol reduction that the United States could achieve by retiring half of its "dirty" coal-fired plants: close to one-third

- Economic losses from weather-related natural disasters in 1998: $89 billion

  Increase in economic losses from previous record year, 1996: 50 percent

- Reduction in U.S. energy consumption needed to reach Kyoto accord: 35 percent

  State that has already achieved this reduction: California

- Average Swede's yearly carbon dioxide production: 1.5 tons and falling

- Average U.S. resident's yearly carbon dioxide production: 5 tons and rising

- Percentage of Kyoto reductions the United States could achieve through economically viable increases in efficiency standards for new buildings: about 8 percent

  Amount building owners would save as a result of the standards: $65 billion

- Percentage of Americans who believe global warming is happening now: 57 percent

  Percentage who believe we should take action without waiting for complete scientific consensus: 66 percent

- Percentage of Americans who say the United States should take the lead in reducing global warming: 75 percent

- Number of cars on the planet in 1950: 50 million

- Number of cars on the planet today: 500 million.

- Fraction of the world's fleet of cars owned by U.S. residents: more than one-third

- Percentage of Kyoto reductions the United States could achieve by eliminating the loophole exempting SUVs and light trucks from fuel-efficiency standards: about 10 percent

- Decrease in cost of photovoltaic cells since the 1970s: 95 percent

- Growth in sales of photovoltaic cells from 1996 to 1997: 43 percent

Source: Ross Gelbspan, "Changing the Climate: The Global Warming Crisis," *Yes! A Journal of Positive Futures,* issue 12 (winter 1999)

4

Vanishing Resources

[This] has been a decade when the darkening cloud of pollution seriously began degrading the thin envelope of air surrounding the globe; when pesticides and unrestricted waste disposal threatened the productivity of all the oceans of the world; when virtually every lake, river and watershed in America began to show the distressing symptoms of being overloaded with polluting materials.

—1970

In our race toward progress it seems we're always in a rush. There was the gold rush that helped push the nation's growth to the Pacific Ocean. The rise of the industrial and automotive age fueled an oil rush that continues to this day. Black gold, they call it, shouting "Eureka!" every time their rigs and drills tap into a new source of oil. We've deployed troops and gone to war to protect our access to Middle East petroleum, a resource of the Earth that, though finite, is a powerful sustainer of a chugging economy. But we must not forget that everything has a limit—every resource, from our forests, farmlands, and grasslands to the rocks filled with minerals to our rushing rivers.

As botanist Peter Raven, the head of the American Association for the Advancement of Science, put it so well, "We ought to be able to change our 'flat world' view of Earth, in which we see the horizon stretching out before us infinitely. We need to see the world as it was photographed in the Apollo space mission—an actual planet that has actual limits."[1]

That flat-world thinking has led the world's population to double its freshwater consumption since 1960, quintuple its burning of fossil fuels since 1950, and increase wood consumption by 40 percent in twenty-five years, as the United Nations reported in 1999. At the same time, we continue to pollute the air and water, erode the soil, and deplete the groundwater. Sometime soon we must pause to take stock of our vanishing resources, while there is still something to save.

Clearing the Air

Any of us who have fought to catch our breath, or watched an asthmatic child gasp for air, know the necessity of clean air. It is as vital to survival as the rhythmic pulsing of our hearts, as fundamental to life as food and water. We draw a breath every 4 seconds, 16 times a minute, 960 times an hour, and nearly 8.5 million times a year.[2]

Most of us don't think about it. Fortunately, breathing is something we take for granted. Yet a growing number of people—especially children and the elderly—find breathing a challenge, at least some of the time. If it seems that more kids on the block are suffering from asthma attacks, or that more of your friends and colleagues are packing inhalers these days, they probably are. Medical experts report that 5 million children in the United States have asthma, and that childhood asthma attack rates have more than doubled in the past decade. Adult-onset asthma is also becoming more common.[3]

What about smog? We've all heard of—or experienced—ozone action days in Los Angeles and some of our other large cities. On those days, when hot weather bakes and worsens ground-level ozone, or smog, those most at risk for respiratory problems are advised to stay indoors until the air clears. But in many other places across the country, smog levels are higher than is considered healthy, even if alerts aren't called. During the summer of 1999, for example, the U.S. Public Interest Research Group reported that smog levels exceeded federal standards in forty-three states, with more than 7,600 recorded violations of federal Environmental Protection Agency (EPA) healthy air standards during an eight-hour period. The same report stated that smog pollution is responsible for more than 6 million asthma attacks and sends 159,000 people to the emergency room each year.[4]

These problems may seem counterintuitive in light of the substantial progress that has been made in the nation's air quality since passage of a retooled Clean Air Act in 1970 and the Clean Air Act Amendments of 1977 and 1990. We, indeed, are breathing easier today than we did back then, owing to strong public pressure on federal lawmakers to restrict what is released into the skies above our communities.

The EPA reported that in the last three decades, air emissions have been cut by nearly a third and acid rain has been reduced in the range of 10 to 25 percent. Emissions of five of the six worst air pollutants—carbon monoxide, lead, ground-level ozone, particulates, and sulfur dioxide—dropped from 1970 to 1999. Most strikingly, a mandated reduction of the amount of lead in gasoline has resulted in the near elimination of lead from car emissions, and lead levels in the air have fallen more than 90 percent in the last twenty years. At the street level, where people are exposed, levels of ozone and carbon monoxide also have improved greatly thanks to improvements in vehicle and industrial technology. The EPA reports that nationally, air quality in the 1990s was better than in any year during the 1980s, showing a trend of improvement.

Notably, these improvements occurred despite a 31 percent increase in the U.S. population, a 114 percent rise in productivity, and a 127 percent jump in the number of vehicle miles traveled by Americans.[5] The statistics are interesting because they tell a far different story than the gloomy picture of a crippled national economy and widespread job loss forecast by industry and others thirty years ago in the face of tightening air quality controls.

Yet, despite this progress in reducing the total tonnage of air emissions, people are still exposed to unsafe levels. The EPA reported that in 2000 about 121 million people in the United States were living in counties where air quality fell short of national air quality standards.[6] Behind those numbers are people: the parent who fears another asthma attack when her child heads outdoors to play, or the elderly person who stays indoors on a sweltering summer day because he has difficulty breathing. Why do large swaths of the country continue to experience substandard air quality fully thirty years after a more stringent Clean Air Act went on the books and more than a decade after the Clean Air Act amendments of 1990?

One reason is that many older and dirtier power plants that operated in the early 1970s have been exempted or grandfathered. Another is the problem of extremely fine particulates in the air—the sort that until relatively recently had been considered too small and insignificant to be subject to federal regulation. Finer than a human hair, these particulates escape the bag houses and scrubbers used to trap pollutants from industrial smokestacks. Once airborne, they can make their way past our own protective breathing apparatus and deep into the most sensitive areas of our lungs, lodging there and causing respiratory problems. Although the American Lung Association, the American Public Health Association, the American Academy of Pediatrics, and the Asthma and Allergy Foundation of America have called on the EPA to regulate fine particulates, corporate lawsuits and lobbyists have delayed action by the agency.[7]

The statistics also don't show that emissions of nitrogen oxides have increased 17 percent since 1970. These emissions contribute to the formation of ground-level ozone, which dims the skies and can damage the lungs, aggravate respiratory problems, and irritate eyes. Nearly one hundred U.S. cities are reported to have unhealthy ozone levels, and the EPA considers nine of the cities—home to 57 million people—severely polluted.[8]

Smog has other effects as well. For example, each year, ground-level ozone reduces agricultural and commercial forest yields in the United States by more than $500 million. In China, the world's fastest-developing country, smog is damaging crops, raising questions about whether that country will be able to adequately feed its growing population.[9]

Smog also destroys the beauty of many of our most treasured lands and vistas. Haze obscures much of the scenery in the Great Smoky Mountains National Park, where smog concentrations in remote areas of the park have increased nearly 20 percent over the last ten years. Throughout the eastern United States, regional haze has reduced average visual range to just 15 to 30 miles—about one-third of the range typical of natural conditions. In Shenandoah National Park, located within the Blue Ridge Mountains and once known for its outstanding vistas, today you can see 20 miles or so where you once could see 90.

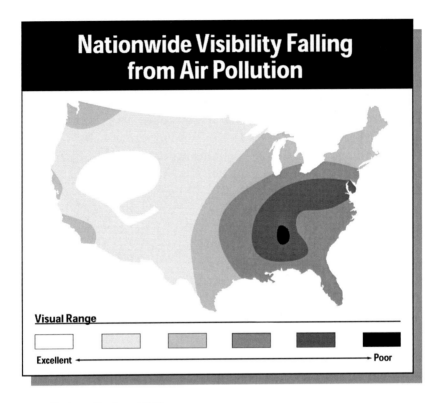

Air pollution affecting visibility.
Since 1970, the United States has reduced some types of air pollution. But visible and invisible pollutants continue to shorten life spans and exacerbate asthma and other lung diseases. The air's moisture content affects visibility, but even after taking humidity into effect, visibility is still dropping over large parts of the nation. The cleanest air in the lower forty-eight states is in the Rocky Mountains; the haziest is in the Southeast. Regional haze has reduced annual average visibility in these areas to about one third of their natural levels in the West, and to one quarter of their natural levels in the East. *Illustration:* Chuck Lacasse of Resurgent Creative, Green Bay.
Source: Clean Air Task Force, 2001.

"People talk about how hazy it is, but they don't realize what's causing it," said Cindy Huber, U.S. Forest Service air resource specialist for the southeast region. "Some are conditioned to it and some think it's natural because, after all, it's the Blue Ridge Mountains or the Great Smoky Mountains and, as their names imply, that's how they're supposed to be. But that's not true. What we're seeing is far from natural, far from how these mountains are supposed to be."[10]

The problem isn't confined to the eastern United States, though the loss in visibility is greatest there. Poor visibility at the Grand Canyon has been front-page news for a number of years; I have been at the South Rim on days when one could not see across the canyon. Similarly, stand near the crest of Mount Rainier in Washington's Cascade Mountains on a warm summer afternoon, and you're likely to see only traces of the vast Pacific mountain chain that stretches far into the distance. On some days the views are best from inside a cliffside kiosk, where pictures of the mountain range on clear and hazy days educate the public on the smothering effects of smoggy ground-level ozone.

It's not only daylight views that are obscured. Rangers at some of our national parks complain of not seeing the stars at night and worry that upcoming generations may miss out on the spectacular starlit skies so many of us have taken for granted.

Cars and trucks are a major source of the air pollutants that cause smog and other environmental problems, and vehicles are the single greatest polluter in some of our major cities. In fact, the most polluting thing many of us do each day is drive our cars. Every time we start the engine, a cocktail of nitrogen oxides, carbon monoxide, hydrocarbons, and carbon dioxide spews out the tailpipe—contributing to smog, acid rain, and global climate change.[11]

We can and always have been able to build cleaner cars, but progress is in spurts—a great leap forward, then resistance. First, the successes: since 1970, efforts by the government, environmentalists, and industry have substantially curbed vehicle emissions. Not the least of those efforts was the first-generation catalytic converter in 1975, introduced to reduce polluting hydrocarbon and carbon monoxide emissions. This effort to clear the air also provided a secondary benefit: because lead renders the catalyst inactive, the mid-1970s saw widespread introduction of unleaded gasoline. That, followed by a ban on leaded gasoline in 1986, led to a 90 percent reduction in the amount of lead in the air over the United States between 1977 and 1987 and a corresponding decrease in lead levels in people's blood. This reduction was especially critical for children, whose developing brains and nervous systems are especially vulnerable to damage from lead—

damage that can result in learning disabilities, hearing loss, behavioral abnormalities, and a drop of several points in IQ. In the 1970s, health surveys found nine in every eleven American children had lead levels in their blood that measured above the threshold for safety. Fortunately, research shows that average lead levels in children's blood have dropped 75 percent. Yet one in every eleven children still was exposed to dangerously high levels of lead as recently as the late 1980s.[12]

In the case of the catalytic converter, an effort to stop one pollutant helped curb others. Yet many of the small steps toward progress in curbing vehicle emissions have been offset by the fact that the number of cars and number of miles driven in the United States during the last thirty years have about doubled. More than 215 million passenger cars and light trucks travel U.S. roads, driving nearly 2.5 trillion miles each year.[13] That isn't expected to slow anytime soon, given the nation's continuing population growth and EPA projections that vehicle travel in this country now doubles every twenty years.[14] A total of 317 million passenger vehicles are projected to be on the road by 2050, according to the Department of Energy.[15] That will put an additional 100 million cars on our highways within the next fifty years. These steep increases in vehicles and vehicle travel will come at great environmental cost if they continue to outpace gains in emissions technology. Moreover, where will we put another 100 million cars?

A doubling in vehicle travel could also bring about a doubling of air pollutants that stream out of the tailpipe. Unless we dramatically reduce air pollution from vehicles or drastically cut back the amount we drive, smog-free air will continue to elude us, and this major contribution to global climate change and acid rain will worsen.

Yet there are all kinds of things we can do to reduce carbon dioxide and other vehicle emissions if we put our minds to it and if the public understands the need to do so. The technology already is here to double the nation's Corporate Average Fuel Economy (CAFE) standards, which require auto makers to achieve 27.5 miles per gallon for cars and 20.7 mpg for light trucks.

A couple of years ago I rode in a Japanese hybrid. It was a comfortable car that got 60 miles per gallon, ran on gasoline and batteries, and hummed along quietly. We should be aiming for 80 miles per gallon.

It is also time that we bring gas-guzzling sport-utility vehicles (SUVs) and light trucks into the twenty-first century by requiring that they conform to the same fuel economy standards that cars have met for years. Why don't they? SUVs, minivans, and light trucks benefit from a legislative loophole that allows all vehicles defined as "light trucks" to meet less-stringent fuel economy standards than passenger vehicles. The provision dates back to the 1970s, when light trucks were fewer and were operated primarily by farmers and businesses. But auto dealers soon developed a market for these roomier passenger vehicles, and by 2001 they accounted for more than half the new vehicles sold in the United States.[16] Out on the road, 35 percent of passenger vehicles—or about 71 million vehicles—were classified as light trucks or SUVs in 1998.[17] Some of the big automakers are selling even larger versions today.

Repeated attempts to pressure Congress to close this loophole have failed. Yet accomplishing this is in the nation's long-term interest. A July 2001 National Academy of Sciences (NAS) study said as much. In it, the nation's top engineers reported that available technologies could make SUVs more fuel-efficient and cleaner, with no long-term economic impact to SUV buyers.[18]

Everybody bears the environmental cost of this loophole. SUVs and light trucks pollute the air, affecting our climate, plants, fish, and soils—not to mention public health. Granting special treatment for these vehicles translates to burning more gasoline and releasing more pollutants into the air every year it remains in effect—something we don't need, considering the vast amount of pollutants emanating from tailpipes of all types already.

Each gallon of gas burned, for example, sends nearly 20 pounds of carbon dioxide—a principal greenhouse gas—into the atmosphere.[19] We know that the average car releases 6 tons of carbon dioxide into the air each year, and the average SUV or light truck releases about 10 tons. Multiply this tonnage by the more than 215 million vehicles on the road today, and that's more than 1.5 billion tons of carbon dioxide released into the air each year, and more than 15 billion tons released over a decade.[20]

The numbers are staggering. It's time to toughen the CAFE standards for vehicles classified as light trucks as well as for standard cars.

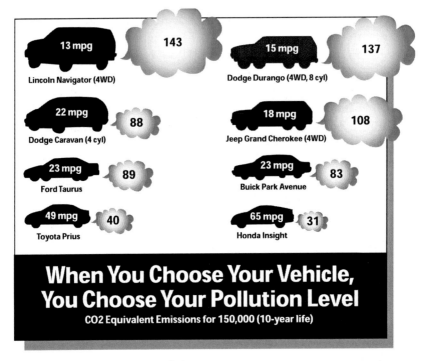

When You Choose Your Vehicle, You Choose Your Pollution Level
CO2 Equivalent Emissions for 150,000 (10-year life)

Consumer choices, government choices.
A federal loophole allows automakers to sell a single, gas-guzzling SUV in 2002 that emits as much carbon dioxide as four to five highly fuel-efficient autos. The technology exists for automakers to manufacture SUVs, light trucks, and minivans that are as clean and efficient as high-efficiency cars if the government were to hold these large vehicles to the same standards as automobiles. *Illustration:* Chuck Lacasse of Resurgent Creative, Green Bay.
Source: National Research Council, *Effectiveness and Impact of Corporate Average Fuel Economy (CAFE) Standards* (Washington, D.C.: National Academy Press, 2001); National Academies, "Federal Fuel Economy Standards Program Should Be Retooled," <www4.nationalacademies.org/news.nsf/isbn/0309076013?OpenDocument> (accessed April 15, 2002).

Experts tell us the technology is here to significantly improve the fuel efficiency of both types of vehicles within the next fifteen years, depending upon future regulatory requirements and development cycles in this country. Automakers already offer many of these improvements in overseas markets such as Japan and Europe. Still, it is the SUV, minivan, and pickup truck that offer the greatest potential to reduce our fuel consumption.[21]

The way to get there is similar to the way we increased fuel economy standards from 14.2 miles per gallon back in 1975: we passed a law mandating the new, tougher standards. It worked. U.S. gas consumption is down about 33 percent from what it would be if the current CAFE standards were not in place. Looked at another way, the amount of gas that CAFE saves each year could fuel all the cars, light trucks, SUVs, and other gas-powered vehicles in the four most populous states: California, New York, Florida, and Texas.[22]

There's Something in the Water

As someone who grew up among the many lakes and streams of northwestern Wisconsin, I know well the value of clean water. We use it for boating, fishing, and drinking. We swim in it, bathe in it, build houses around it, and often simply gaze at it.

Water also holds a special place in our memories. Many of us remember family times or vacations spent at the beach, at the cottage, or visiting a prominent local river or waterfall on a class outing. Water was the organizing principle. Take it away, and gone is the magic that drew us there in the first place.

Much of my youth was spent exploring the marshes and shorelines bordering Little Clear Lake, Big Clear Lake, and Mud Lake—three small bodies of water that anchor my hometown at both ends of Main Street. The town itself took the name Clear Lake, a practice repeated across this nation, around the world, and throughout history as a reflection of the importance water plays in defining our sense of place and identity. Unfortunately, our reverence for water doesn't protect it from abuse. During the Great Depression of the 1930s, a Works Progress Administration project drained most of Mud Lake and built a road across the lake bed to the east end of Main Street—saving about 30 seconds of time traveling from South Street to Main Street. Big deal! That cost us several muskrat houses and a good landing and resting spot for migrating ducks.

Nearly three-quarters of a century later, the story is no different—we continue to degrade our waters and render them unusable, both for ourselves and for future generations. Clearly, the quality of much of the nation's waters has come a long way from the days when the

Cuyahoga River burned and Lake Erie was proclaimed dead. The Federal Water Pollution Control Act (FWPCA) amendments of 1972 are responsible for helping us turn the page on a period of this nation's history when our rivers, lakes, and streams were dumping grounds for our wastes. We learned from the experience, however, that in polluting the water we fouled our own habitat. Popular gathering places were closed off to beachgoers and anglers. Waters that once teemed with life became killing fields for fish and other aquatics. Tranquil resting spots were overtaken by the rank odors of sewage and rotting fish. And it wasn't long before entire cities turned their backs on their waterfronts and focused their attentions inland.

The general goal of the FWPCA, amended in 1977 to become what is now known as the Clean Water Act, was to "restore and maintain the chemical, physical and biological integrity of our nation's waters." That has happened, to some degree.

In the Chesapeake Bay, for example, water quality problems persist, but the bay is on the upswing overall. Bay grasses have increased 60 percent since 1984, striped bass populations have reached historically high levels, and wild shad are increasing in number. Most of the bay's major tributaries are running much cleaner than a decade ago, and chemical releases in the bay watershed have shown a 55 percent drop from 1988 to 1994.[23]

Yet, thirty years after the public made it clear to lawmakers that clean water was a national priority, many of our waters are still troubled. How many of us, for example, live near a lake, river, or freshwater bay that is too dirty to swim in, too polluted to tap as a water supply? How many of us have seen the fish return, only to find we can't eat them because they're contaminated from chemicals?

In the EPA's most recent reports to Congress, the agency noted that although numerous improvements have been made to the nation's water quality, "serious water pollution problems persist nationwide." According to the EPA, about 40 percent of the nation's surveyed rivers, lakes, and estuaries are too polluted for basic uses such as fishing and swimming.[24]

In short, we're still far wide of the mark we set for ourselves. Yet many people are unaware of this. Why? In part, because state and local governments have little or no monitoring in place for public waters. Environmentalists have called attention to this issue, and their efforts have prompted more monitoring today than existed a decade ago.

But where monitoring is conducted, water quality tests find many beaches unfit for swimming. In 2000, for example, the number of beach closings forced by bacterial pollution—such as *E. coli* bacteria—doubled across the nation from the year before. The doubling was attributed chiefly to more monitoring, better testing standards for bacteria and other pathogens, and more complete reporting. Analyzing data collected by the EPA, the Natural Resources Defense Council (NRDC) reported 11,270 beach closings nationally in 2000, up from 6,160 the prior year. Polluted storm water runoff was cited as the leading source of the bacteria, followed by pipeline breaks or sewage treatment plant failures.

The NRDC, which has been monitoring beach health for 11 years, reported another disturbing trend: the number of beaches reporting pollution problems from an unknown source rose from 40 percent in 1999 to 56 percent in 2000. "We're seeing a much more realistic picture of the beach water pollution problem now that more states are monitoring and reporting, but we haven't turned the corner on identifying the sources of pollution and preventing them in the first place," said Sarah Chasis, an NRDC senior attorney and director of the organization's water and coastal program. "It's outrageous that more than half of the time local authorities didn't know where all the pollution was coming from when they had to close a beach or post an advisory."[25]

Lake Michigan beach closings are commonplace in the Milwaukee area, though health officials there are uncertain of the source or sources of bacterial contamination. Seagulls, storm water runoff, and pet wastes are all suspected. Another suspect is the Milwaukee Metropolitan Sewage District (MMSD), which has dumped more than 13 billion gallons of raw sewage into Lake Michigan and its local tributaries since 1994 because of inadequate capacity to temporarily store storm water overflow.[26]

Such continued sewage dumping galls those who believe the nation's waterways should be better protected under the federal Clean

Water Act of 1972. "The fact that raw sewage is still being dumped into our nation's waterways 30 years and beyond after the Clean Water Act is appalling," said Cameron Davis, executive director of the activist Lake Michigan Federation, which is suing the MMSD for sewage dumping. "A lot of people thought that once a law was on the books, your problem was solved—but that couldn't be further from the truth. Good laws require good implementation."[27]

If you have a polluted river running behind your home or through the middle of your town, you're not alone. The overwhelming majority of Americans—218 million—live within ten miles of a polluted lake, river, stream, or coastal area. States have identified almost 300,000 miles of rivers and streams and more than 5 million acres of lakes that don't meet state water quality goals and therefore are considered unsafe for swimming or cannot support healthy fish and other aquatic life.[28]

If we step back and take a general look, it's understandable that many believe water pollution in the United States is under control. How many have the time or ability to check? Don't all cities have wastewater treatment plants? Isn't industry complaining about the onerous expense of controlling pollution?

Thirty years ago the government targeted the "discharge pipe"—the pipes industries use to discharge wastes directly into nearby waters. The government crackdown substantially cut industrial water pollution, bringing about a change so dramatic that rivers that once had nary a fish worth catching now do.

Not all rivers or estuaries were restored to health, however. The chief culprit is runoff—the soil, chemicals, and other materials that run off the surface of the land with each rainfall and snowmelt. This mix inevitably beats a path to the nearest river, lake, or stream, introducing a brew of foreign substances ranging from pesticides and fertilizers to road salt and gasoline. Each pollutant then goes to work altering the delicate ecology of the waterway and affecting the critters that dwell therein. Phosphorus and nitrogen from nutrient-rich farm fertilizers and manure dissolve and promote the growth of algae and other vegetation. Excessive algae growth, in turn, alters the aquatic environment by blocking sunlight and sucking up oxygen

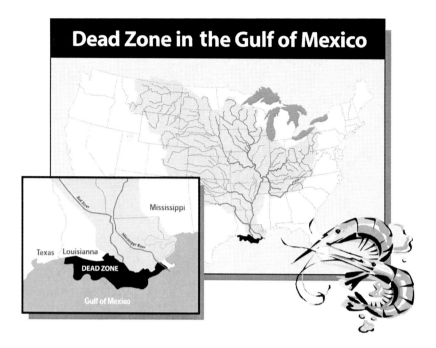

Dead Zone in the Gulf of Mexico

Fertilizers are damaging coastal fisheries.
Fertilizers are deluging the Mississippi River, creating a vast "dead zone" off the Louisiana-Texas coast. Excessive fertilizer use and poor erosion control in states draining to the Mississippi are lowering oxygen levels on the ocean floor in the Gulf of Mexico and killing bottom-dwelling organisms. The dead zone exists most of the year amid the most important commercial and recreational fisheries in the United States, threatening the region's economy. *Illustration:* Chuck Lacasse of Resurgent Creative, Green Bay.
Source: National Research Council, "Clean Coastal Waters: Understanding and Reducing the Effects of Nutrient Pollution," Committee on the Causes and Management of Eutrophication, 2000.

when the plants die—lowering oxygen levels that fish and other organisms depend upon.

This is no small matter. We have low-oxygen and no-oxygen conditions in more than half of the 136 major estuaries along the coasts of the lower forty-eight states—places such as San Diego Bay, Long Island Sound, and Skagit Bay, Washington. If these conditions persist year after year, a "dead zone" of lifeless bottom waters develops. Spawning areas are ruined, and a habitat and food chain disaster cascades through the region. Off the Louisiana and Texas coasts, one of these ocean dead zones spreads more than seven thousand square miles, an area the size of New Jersey.[29]

Another contributor to runoff pollution is the loss of trees, plants, and tall grasses alongside rivers and streams. Marshes and wetlands once guarded the perimeters of many lakes and waterways, holding back soil, nutrients, and contaminants. But many of these stalwart soldiers have lost their footing to bulldozers, backhoes, and other tools

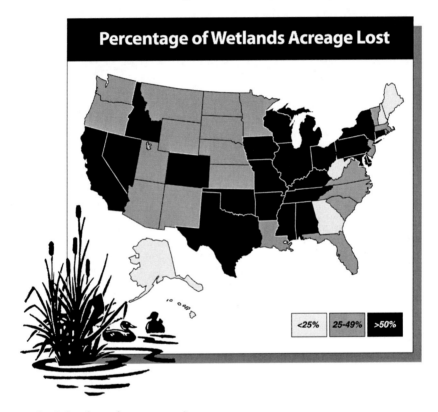

Percentage of Wetlands Acreage Lost

<25%   25-49%   >50%

Wetlands loss hurts future generations.
The United States has seen many of its rich wetlands disappear, habitat that was home to a superabundance of birds, fish, plants, and animals. Since the nation's founding, twenty-two states have lost at least half their original wetlands to draining and filling. Seven of these states (California, Indiana, Illinois, Iowa, Missouri, Kentucky, and Ohio) have lost more than 80 percent. Wetland losses translate to other losses as well, depriving current and future generations of cleaner streams, rivers, and lakes and adequate recharge of groundwater supplies. *Illustration:* Chuck Lacasse of Resurgent Creative, Green Bay.

of development and progress. The toll is beyond estimation, and this is one reason our ocean fisheries are in serious trouble. The U.S. Fish and Wildlife Service reports that between the 1780s and 1980s, twenty-two states lost at least 50 percent of their original wetlands—with seven states losing more than 80 percent.[30]

Unlike the pollution of the past, the sources of these pollutants can be hard to identify. After years of dealing with the so-called point source pollution from the discharge pipe, we now must contend with these nonpoint sources—so named because they are so varied and diffuse.

Scientists know full well that agriculture is the leading source of runoff pollution from nutrients, pesticides, and silt. Farm fields bleed dirt, manure, fertilizers, and pesticides into our surface and ground-water during heavy rains. Though farmers as a group are relatively easy to identify as chief sources of these pollutants, individually they often lack the financial means of big industry to change their practices. For these reasons, coupled with strong opposition from the agricultural lobby, the federal government has shied away from imposing stringent runoff restrictions on farmers.[31]

But farm fields aren't the only sources of runoff. Urban streets, sidewalks, parking lots, and driveways also contribute a significant share of pollutants to waterways. Think, for a moment, of your own property during a rainstorm. The rain first hits the roof of your house and garage, bouncing off with greater force than if it had struck grass or plants. From there it falls to the paved driveway, where it picks up oil and gas drippings from the car. The rain mixes with the fertilizer applied to the lawn and the pesticides applied to the flower garden. It gains speed as it cascades over the sidewalk, sweeping up grass clippings, loose soil, and pet wastes on the boulevard before making its way to the storm sewer by the curb. From there—well, Maryland school children have painted warnings by curbside storm drains in their towns that read "Save the Bay"—even though the Chesapeake is more than fifty miles away.

The kids are right. Every time we overapply fertilizer and pesticides to the grass and flowers, every time we hose down the driveway, and every time we use salt to melt the ice on the sidewalk, the residue flows eventually to the nearest waterway. No one likes to look in the mirror and think of himself or herself as a polluter, but all of us need to realize we have a hand in dirtying the water.

Why, some might ask, should we care about what goes into lakes and rivers as long as clean water flows out when we turn on the tap? Understand that water pollution is not only a threat to wildlife and recreation, but it can affect the water in the tap without our even knowing it.

Sometimes the effects are dramatic. In 1993 in the Milwaukee area, more than 403,000 people fell ill and more than 100 people died from cryptosporidiosis, a gastrointestinal illness believed to have resulted when the drinking water supply was contaminated with an intestinal parasite from area livestock. A follow-up study by the U.S. Centers for Disease Control and Prevention in 1997 speculated that human sewage might have been the real culprit. Whatever the cause, it was the nation's single largest waterborne epidemic.[32]

Sometimes the effects are not so obvious. Fish from polluted waters often are unsafe to eat because they absorb chemicals that have slow but serious effects on human health. Across the nation, more than 2,500 lakes have fish consumption advisories that warn anglers to limit their consumption of certain fish or not eat them at all.

In the Great Lakes, home to 20 percent of the world's surface freshwater and around which more than 25 million Americans live, more than 90 percent of the U.S. shoreline is under fish consumption advisories. Public health officials say the greatest concern is the fishes' accumulation of PCBs, or polychlorinated biphenyls, poisonous chemicals discharged in large quantities by industry before production was banned in the 1970s.[33]

Mercury, which disrupts the brain and nervous system of the developing fetus and in both children and adults, also contaminates fish and seafood—which is our greatest source of exposure to this metallic element. Forty states have issued advisories restricting consumption of freshwater fish containing elevated levels of methylmercury, a form of mercury that is easily absorbed by our bodies.

The National Academy of Sciences reported in 2000 that more than 60,000 newborns a year might be at risk for developmental problems of the nervous system, resulting from exposure to elevated mercury levels in the womb of mothers who eat tainted fish. Research shows these deficits are likely to lead to poorer school performance for the affected children, meaning they may struggle to keep up in a normal classroom or may require remedial classes or special education.[34] Another study, this one released by the U.S. Centers for Disease Control (CDC) in 2001, found that about 10 percent of American women of childbearing age have mercury levels within one-tenth of potentially hazardous levels—a narrow margin of safety that underscores the need to reduce methylmercury exposure. The CDC study is notable because it was the first human tissue measure of the U.S. population's exposure to mercury.[35]

Some folks will note that mercury is a heavy metal that occurs naturally in the environment, but our society also releases great quantities of it into the skies. Utilities that burn coal to produce power contribute the largest share, releasing approximately forty tons annually in the United States; municipal and medical incinerators are the next largest human generators of mercury pollution. Once released into the air, mercury returns to the Earth in rain, where it ends up in lakes and waterways and is converted by bacteria into a form that easily works its way up the food chain—from tiny bottom-feeding critters to fish to birds and people.[36]

A river or lake too murky to reflect our gaze may, indeed, be the truest reflection of our impact.

Our sullied lakes, rivers, and streams are evidence of the toll we are taking on the planet as our numbers and demands increase. The burden is too great when the soil runs off the land unimpeded and clouds the waters. The land is overtaxed when fertilizers and manure make water unfit for swimming or drinking. We fool no one but ourselves in presuming that the deadly chemicals we flush into the water will go away and leave the underwater world intact and unaltered. And we set ourselves up for greater problems as we increase the pace with which we pave over the Earth's surface—laying down acre upon acre of new roads, sidewalks, parking lots, shopping centers, office parks, and homes.

We can take heart from the fact that more than half of the nation's rivers, streams, and lakes are improving. Many have been restored in just the last three decades, thanks to a watchful and wary public. Know, however, that the easiest projects—those providing the greatest bang for the buck—have largely been implemented. We're more than halfway to our goal, yes, but we're not all the way.

Water, Water Everywhere?

The astronauts' view from space as they look down on Earth is mostly blue, with oceans outnumbering lands by an overwhelming ratio. The truth is, however, that though the world may be overflowing in the saltwater of its oceans, it is dying of thirst for what is becoming the most precious of all resources—freshwater. "The wars of the next century will be about water," summed up World Bank vice president Ismail Serageldin.[37]

Water—which some now call blue gold—is becoming a precious commodity as worldwide consumption of this finite resource increases. Humanity now uses more than half of the available surface freshwater on Earth. Although the world's supply of freshwater remains constant, our consumption of this most vital resource doubles every twenty years—more than twice the rate of human population growth.[38]

In the growing developing nations, annual water demands by households and industries will climb by 767 billion cubic yards between 1995 and 2020, a volume equivalent to the annual flow of seven Nile Rivers, according to the International Food Policy Research Institute.[39]

Don Hinrichsen of the Johns Hopkins University School of Public Health wrote in a 1998 population report, "In many developing countries, lack of water could cap future improvements in the quality of life. Meanwhile, there is no more freshwater on Earth than there was 2,000 years ago, when population was three percent of its current size."[40]

Around the world, water scarcity has played a role in dividing the haves and have-nots, prompting migrations that strain already overpopulated cities and contributing to border and territory disputes among nations.

Water supply is not just an issue for poor developing countries, however. Parts of the United States join areas, of China, India, North Africa, and the Middle East in facing the most severe groundwater depletion problems.[41] Here in the United States, the issue will take on even more importance during the next twenty years. As we aim to meet the demands of a growing populace, water tables are falling. Farmers working to expand production and imports are drilling more wells for irrigation—and, in turn, are mining groundwater aquifers dry.

More and more of our population is moving to arid places such as Arizona, Colorado, and the deserts of California, and they want the same green lawns they have in wetter parts of the country. But it's not only dry areas that face water shortages. Even rain-plagued Seattle sees a shortage looming on the horizon as population numbers swell there.[42]

In addition to the effects on people, heavy water demand can harm fish and wildlife. Around Washington's soggy South Puget Sound, which receives more than fifty inches of rain each year, it is feared that more and more groundwater withdrawals by a growing population will reduce water flows to area waterways—threatening the endangered salmon and other fish.[43] The Colorado River delta, so important for bird and fish reproduction, is desiccating and shrinking because of upriver water withdrawals. The renowned winter bird sanctuary of the Florida Everglades has been damaged by water withdrawals to supply the cities of South Florida. Across the nation, many lakes and streams are dependent on aquifer flows and the normally cooler temperatures of groundwater, which holds more oxygen for fish and other aquatic life.

Draining the ground's freshwater supplies will increase our reliance on desalination plants to tap the world's oceans. But experts caution that we have no solution for what to do with the mountains of salt created through desalination.

Salt already is contaminating groundwater where freshwater is withdrawn at a rate beyond the capacity of aquifers to naturally recharge. From Southern California's Orange County to South Carolina's Hilton Head Island to Long Island, New York, heavy pumping of aquifers over decades is drawing saltwater into the drinking water aquifers, making the water unusable.[44]

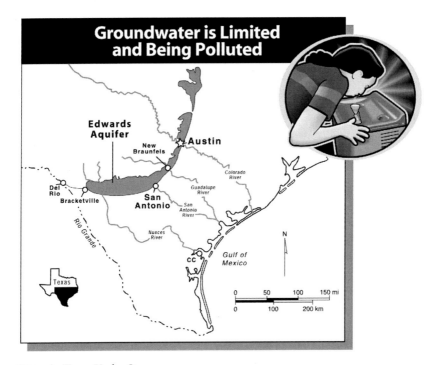

Water in Texas Under Stress.
Like many regions of the United States, south central Texas depends on underground aquifers for water. In aquifers such as the Edwards Aquifer, withdrawals are approaching recharge levels. The Edwards Aquifer supplies water not only for people but for local streams near Austin and San Antonio. Social conflicts are predicted as the state's population of 21 million is expected to double in fifty years, with the water supply already stretched thin. *Illustration:* Chuck Lacasse of Resurgent Creative, Green Bay.
Source: John M. Sharp Jr. and Jay Banner, "Water for Thirsty Texans," in *The Earth around Us* (New York: W. H. Freeman, 2000).

Other water sources in the United States are threatened as well. In parts of the Texas Panhandle and New Mexico, a giant underground aquifer known as the Ogallala is being depleted. This groundwater source is so big it underlies parts of eight states: Texas, New Mexico, Nebraska, Oklahoma, South Dakota, Colorado, Kansas, and Wyoming. Yet irrigation in the Texas High Plains and surrounding states is depleting the aquifer at a rate of 12 billion cubic meters per year, causing some farmers to go back to wheat and other less valuable dry land crops.[45] The Edwards Aquifer—water source to areas of Texas near Austin and San Antonio—is another example of an aquifer under severe pressure from a growing population.

As pressures on clean water continue to grow in the face of scarcity around the world, water users are looking to other freshwater sources. Residents of the Great Lakes region already worry about companies eyeing its vast supply of freshwater. Those concerns were justified in the late 1990s when an Ontario company sought to sell water from Lake Superior and ship it to the Far East. The Ontario government granted a permit for the withdrawal but reversed its decision after pressure from U.S. and Canadian authorities. Tapping the Great Lakes may not be an imminent threat to the region's ecosystem, but it is cause for concern. The International Joint Commission, which includes representatives of both nations' governments, has studied whether water can be spared from the basin without disrupting its delicate ecological balance. Based on this work, the commission in 1999 recommended a moratorium on diversions of water from the Great Lakes, citing the potential effects of global climate change and future water demand. Climate-change models from the Great Lakes Environmental Research Laboratory predict that evaporation induced by global warming, for example, could drop the level of Lake Michigan by about three feet in the next thirty years.

Because the Great Lakes region is home to one-fifth of the world's surface freshwater, it is considered a potential target for more populous areas where aquifers are being depleted. Some worry, for instance, that the tremendous population growth in the southwestern United States someday might prompt calls for a canal reaching from the Great Lakes to Arizona. Reg Gilbert of the watchdog group Great Lakes United recalled construction of the giant Hoover Dam on the Colorado River, used to irrigate areas of Southern California and other southwestern states. "It sounds absurd," he said of diverting Great Lakes water to the Southwest. "But the Hoover Dam sounded absurd in 1880, and only fifty years later there it was, changing the ecology of thousands of square miles."[46]

Saving the Rain

Conservation is the greatest, and simplest, tool we have to keep our water supplies intact. Fortunately, some efforts are under way to conserve water and use it more wisely. There are now front-loading washing machines, for example, that use less water, and better farm irrigation methods are used in some places. Some states, such as California and Florida, are installing systems to "reuse" water, filtering and treating wastewater for non-drinking purposes. Some states are enacting laws to promote conservation. Technology is being introduced to collect rainwater and surplus water in underground reservoirs, and only water conserving showerheads and toilets can be sold in the United States.

But more can be done. We can promote conservation by removing the federal subsidies on water supply that feed overuse. If the resource is practically free, there's little incentive for people to use water efficiently. We can also promote the supply side of the equation by doing more to preserve and restore wetlands, often the place where surface water re-enters aquifers.

The average person may think of water use as something that happens in their home. In the United States, however, agriculture is the largest consumer of water—comprising more than 80 percent of withdrawals throughout much of the American West, for example. It is also one of the most inefficient, with much of the water lost to evaporation, leaking pipes, and other factors en route to the farm field. If we are to deal with water scarcity, the most productive step we can take is to improve the efficiency of crop irrigation.[47]

Global water expert Sandra Postel has called on farmers worldwide to rethink irrigation practices, to abandon dams and wasteful big pipes in favor of more sustainable small-scale irrigation techniques. Curbing population growth also can play a major role in protecting our dwindling water supplies. "Coupled with reductions in high-end consumption, slower population growth would make the goal of adequately feeding all people while keeping natural ecosystems intact far more achievable," Postel said.[48]

Fuel from Fossils: Hitting Rock Bottom

Americans are accustomed to using high levels of energy. We use twice as much per capita as industrial nations with comparable standards of living and household income.

Powering American productivity and our lifestyles is a nonrenewable energy system based on oil, natural gas, coal, and uranium. The world's supply of these fossil fuels is finite, we know, but computer modeling and other technological breakthroughs have made it easier for us to find more of them. If there are more sources of fuel to be found, why remind ourselves of the OPEC oil embargo lines at the gasoline station and the government's call to conserve and keep our thermostats low?

Yet these memories of the 1970s energy crisis are still relevant today. Even as the supplies seem less finite than they once did, they are not inexhaustible. Consider oil, arguably one of our most precious resources because of its versatility and value to the nation's economy. The world's oil supply is, indeed, finite. The critical question, however, is when will we exhaust the easy-to-access oil that can be produced cheaply, the type of oil that fueled the world's economic boom from World War II to 2000? Analysts differ on this point considerably. Since the 1950s, most estimates of the world's total oil resource base have ranged from 1.8 trillion to 2.3 trillion barrels. The U.S. Geological Survey (USGS) recently projected a resource base of 3 trillion barrels. Most experts agree that by the mid-1990s, 800 million barrels of that total had been extracted from the Earth.

Based on the latest USGS estimate, the U.S. Energy Information Administration projects that conventional world oil production may increase for another two decades or more before it begins to decline and prices increase.[49] That scenario is optimistic by some accounts, however. Petroleum geologists Colin Campbell and Jean Laherrère, for example, predicted in a 1998 *Scientific American* article that the world's production of oil would peak between 2004 and 2010.

The actual year that peak production occurs, of course, will depend on a variety of factors, among them the growing demand for oil, technological advances, the use of alternative fuels, and the rate of

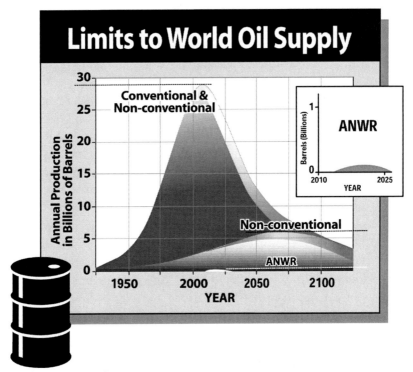

Limits to World Oil Supply

The end of cheap oil is near.

The economic and population growth of the past century has depended on inexpensive "conventional" oil. Inexpensive oil is limited, and almost half the total supply is gone. Sometime in the next ten to twenty years, the world will begin to have less oil available each year—even as the population continues to grow. More expensive "nonconventional" oil will be available in smaller quantities from tar sands, oil shale and the like. The tallest curve shows an estimate of adding nonconventional oil (lighter gray) flow to conventional flow (black). As shown in the inset, if the Arctic National Wildlife Refuge (ANWR) is tapped for oil, it will make little impact in supplying the needed flow to world or U.S. supply. *Illustration:* Chuck Lacasse of Resurgent Creative, Green Bay.

Source: Natural Resources Defense Council, "A Responsible Energy Policy for the Twenty-first Century," New York, 2001, <www.nrdc.org/air/energy/rep/repinx.asp> (accessed April 15, 2002); Kenneth Deffeyes, *Hubbert's Peak: The Impending World Oil Shortage* (Princeton, N.J.: Princeton University Press, 2001).

production. Whenever that peak occurs—whether it is ten or forty years from now—it won't mean the end of oil. What it will mean is escalating prices as the resource becomes more and more precious. As

Campbell and Laherrère put it: "The world is not running out of oil—at least not yet. What our society does face, and soon, is the end of the abundant and cheap oil on which all industrial nations depend."[50]

For those who may have forgotten or are too young to remember, the Arab oil embargo of the early 1970s showed this nation just how oil-dependent our economy is. The U.S. gross national product fell 6 percent between 1973 and 1975, and unemployment doubled to 9 percent.

"The reality is that we've built our civilization around something we knew wasn't going to last," says futurist Richard B. Anderson.[51] And around something quite irreplaceable. Many people may not realize that although fossil fuels burn quickly, they took billions of years to form. It's something to think about. We're burning up nature's efforts in merely a matter of minutes.

Population and growth play a monumental role in the energy problem. Worldwide energy consumption is projected to swell by 60 percent over the next twenty years.[52] Here in the United States we are on the path to doubling our population within this century, and experts tell us that will mean at least doubling our annual energy consumption if we don't take advantage of opportunities to be more energy-efficient. Indeed, continued U.S. population growth means this country will need more energy simply to maintain our current level of use.

The United States is home to 5 percent of the world's people but consumes 25 percent of its oil. The country already imports more than half its oil and a growing portion of its natural gas. With just 3 percent of the world's known oil reserves, a larger population means greater dependency on other countries to supply U.S. energy needs, a dependency that leaves the nation more vulnerable to disruptions. Although there is talk about expanding oil drilling to make America more self-sufficient, the fact is that this country cannot be self-sufficient in oil—unless we reduce our oil consumption by more than half what it is today. Instead, our dependency grows. Oil imports were supplying 57 percent of U.S. needs in 2001, compared to 47 percent a decade before and 36 percent in 1981.[53]

It's a simple equation. In order to lower our dependency we must increase our efforts to be more energy efficient. The potential is there; we must act upon it.

Technological advances are helping renewable energy sources—long criticized as too costly and too far away to be practicable on a large scale—become more affordable and available.

Through most of the 1990s, wind energy was the fastest-growing energy source, followed by solar photovoltaics.[54] Development of a new generation of vehicles powered by hydrogen fuel cells also is underway, with automakers planning to roll out fuel cell–powered autos by 2004. Some predict these vehicles could account for 25 percent of the global vehicle market by 2020. The fact that petroleum giants are investing in solar power is yet another sign of the emergence of renewable power sources.

Still, energy technologies take decades to develop. We must step up our efforts now. We are planning for failure if we wait until 2015 or 2025 to build the new renewable energy system the nation must move to, a system that must provide for nearly 350 million Americans and the world's largest economy.

Conservation is the other key energy challenge we face, and the belief that Americans must lower their standard of living to conserve energy is incorrect. We saw dramatic savings during the 1970s energy crisis, and that was just a small step toward what we're capable of achieving. We can realize huge savings in our energy use if, among other things, we stabilize our population, raise fuel economy standards for vehicles, and improve our energy efficiency.

Rather than using more raw energy to make life better, Americans should work to get more out of the energy they extract and collect. This is true of renewable forms of energy such as solar and wind power, as well as finite resources such as the gas we put in our cars and trucks.

"There are an incredible number of things we can do to live more efficiently," said Peter Raven, of the American Association for the Advancement of Science. "One obvious one is the investment in energy conservation."[55]

Fully two-thirds of this nation's energy goes to business and industry. Fortunately, there has been progress on this front. U.S. manufacturing plants produce more goods with less energy than they did in the late 1970s. Newer office buildings, per square foot of space, use less energy and are more comfortable than older buildings.

Energy savings have been made in many hospitals, clinics, and other health care facilities, for example, through such simple measures as installing high-efficiency reflectors and electronic ballasts in fluorescent lights and tuning up heating and cooling equipment. St. Joseph Hospital in Lancaster, Pennsylvania, has been saving $175,000 a year in electric costs since 1994, when it invested $390,000 to change the type of lighting fixtures in 6,500 lamps. The savings come despite the hospital's doubling its laundry service, opening a 55,000-square-foot surgical facility, and adding new services.[56]

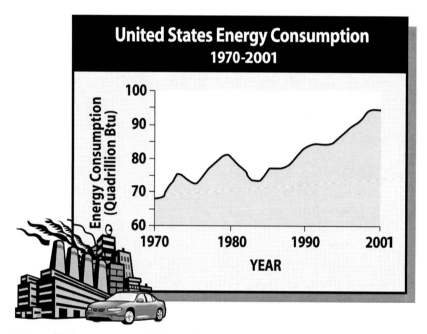

U.S. could choose a more efficient path.
U.S. energy use is growing at a rate that's unsustainable given our dependence on fossil fuels such as oil and coal. From 1970 to 2001 our annual need for energy grew 45 percent, in large part due to a population that grew by 78 million people. If we improved the efficiency of our buildings, factories, and vehicles, the nation's economy could grow until 2010 without using more fuel than it does today—despite continued population growth. *Illustration:* Chuck Lacasse of Resurgent Creative, Green Bay.
Source: U.S. Department of Energy, Interlaboratory Working Group, Scenarios for a Clean Energy Future (Oak Ridge, Tenn.: Oak Ridge National Laboratory and Berkeley, Calif.: Lawrence Berkeley National Laboratory); Statistical Abstract of the United States, 2000.

Actions like this are just a start. As pointed out for years by the Rocky Mountain Institute and the American Council for an Energy-Efficient America, and reinforced by a 2000 study by five federal laboratories, we have barely begun to install the proven energy-saving technologies already at our disposal.[57]

Energy efficiency as a science has improved during the past decade, the result of federal appliance standards developed in consultation with manufacturers. In the next several years, we will see improvements in our water heaters, air conditioners, washing machines, water heaters, and heat pumps.

Recycling is another form of conservation that can improve sustainable living in the United States. Al Gedicks, secretary of the Wisconsin Resources Protection Council, explains the consequence of throwing metal-intensive products out with the trash. "Every time you throw out a color TV, that ends up in the landfill—that's eight pounds of copper that's wasted," he said. "There is a lot more recycling that can be done."[58] Economically speaking, recycling is big business. The worldwide recycling industry now processes more than 600 million tons of materials annually, has annual sales of $160 billion, and employs more than 1.5 million people. In the United States, remanufacturing is already a $53 billion-a-year business and employs some 480,000 people directly—double the number of jobs in the U.S. steel industry.[59]

We can learn from the farsighted policies of other countries, such as those adopted in Europe, Taiwan, and Australia. The European Union is scheduled by 2006–8 to require manufacturers of metal-laden electronic goods—such as computers, televisions, and cell phones—to take back their products and reuse or dispose of them. Already, some manufacturers, from Motorola and Panasonic to Sony and Siemens, are adapting by changing how their products are made, enabling consumer electronic products to be recycled more easily.[60]

Selling the Forests for the Trees

Many of the Earth's forests are being cleared to feed the world's growing appetite for lumber and pulp. Despite heavy logging, regrowth and planting have restored two-thirds of the U.S. Forect Cover. Yet this country faces a different problem as our rich forest canopy is torn asunder piece by piece, posing a major threat to the nation's biodiversity and the survival of sensitive species.[61]

Consider, for example, our shrinking urban forests—woodlands that are not only aesthetically pleasing but contribute an estimated $400 billion in economic benefits through reduced storm-water treatment costs and energy conservation.[62] Unfortunately, vast numbers of trees have been felled by air pollution, disease, and development. Using satellite images, the group American Forests estimates that in greater Seattle the amount of land with more than 50 percent tree-canopy coverage shrank 37 percent between 1972 and 1998. At the same time, the area where the tree canopy measured less than 20 percent more than doubled. On the other side of the country, growth and sprawl helped cut by more than half the heavy tree cover in metropolitan Atlanta—an area once famed for its trees. In the Canton-Akron area of Ohio, heavy tree cover shrank from 55 percent to 38 percent.[63]

A different study found that 160,000 acres of public and private forestland in North Carolina was lost to development and other uses between 1990 and 1999, and that despite tree plantings, the forest experienced a net loss of 15 percent during that same period. The study, by the U.S. Forest Service and North Carolina Division of Forestry, also found that more than half of the Southern Coastal Plain of North Carolina contains only seedling or sapling forests, in which the majority of trees are smaller than five inches in diameter.[64]

The United States is not running out of trees. In fact, the forest lands of the South, Great Lakes, and Northeast have increased somewhat in overall acreage since 1920. From an industrial standpoint, these forests have grown better in terms of sheer pounds of wood added each year than in acreage.

But looking at a forest in terms of acreage, lumber, or wood pulp misses the point. A forest is an ecosystem, and as such, its trees are linked with the soil, soil microorganisms, plants, and animals. The diverse forests of the Central Cascades of Washington, for example, are said to be home to as many as a thousand species of plants and animals. Elk and deer migrate through these forests. Hawks and eagles circle above, salamanders slither below.[65] Many forest species can survive only in large, dense forests having an unbroken canopy of leaves and branches. Cut a road through a forest, and aggressive competitors, predators, and disease agents enter; critical habitat is lost. Woodland creatures once settled deep within the forest suddenly find themselves on a new periphery where they are ill-equipped to live. The same road project thwarts these and other creatures' travel to alternate forestland, blocking corridors once used by wildlife for escape and migration.

Forests have other values that may not be obvious. Consider their role relating to flood control and water supply. Today more than 60 million people in thirty states draw their drinking water from national forest lands. A University of California study quantified the value of the water connection for the Sierra Nevada, for example, and found it more valuable in dollars than the timber supply or anything else.[66]

Recognizing that destroying the resource base destroys jobs, job potential, and profits, many pulp and lumber companies are replanting trees on their land. Other firms also engage in sustainable forestry, a growing movement that aims to protect the forest while harvesting trees. On the retail side, major home supply chains are entering into voluntary agreements to sell only wood that is harvested via ecologically sustainable methods.

Forest ecologists today advocate maintaining the size of certain types of forest, rather than allowing them to be fragmented by roads, checkerboard clear-cutting, and other alterations. The battle to stop forest fragmentation is as controversial as it is critical, however, at times requiring the rerouting of proposed power lines, pipelines, and roads, as well as restricting conventional logging.

This struggle to prevent forest fragmentation is particularly important as we try to maintain the health of the nation's remaining old-growth forests. These ancient forests, which date back thousands of years, maintain an ecological balance that cannot be replaced once lost. Our forests today are relatively young, largely because of the widespread logging of the nation's forests in the late 1800s and early 1900s. Half of the trees in our present forests had not even sprouted when I was first elected to office in 1948. A scant 6 percent are more than 175 years old, defined as old-growth forest.[67]

Whether cleared away, split apart, ground up, or burned, the world's forests are getting smaller. Closed-canopy forests, comprising woodlands where 40 percent of the treetops touch, cover roughly 21 percent of the planet's land area, and only a small number of such forests are protected in the United States and the rest of the world.[68]

Timber from the world's forests is used in a wide variety of products, including napkins and paper towels, fabric softener sheets, disposable diapers, and packaging for everything from soda to dog food. Worldwide demand for paper products is a major contributor to the logging of forests. The wood fiber used to make paper accounts for about one-fifth of the world's total wood harvest, and the world's appetite for paper is growing. By 2050, pulp and paper manufacturing will account for more than half of the world's industrial wood demand.

Global paper consumption continues to climb despite promises that computers and the Internet would create a "paperless-office" economy. As recently as 1997, the world produced 299 million tons of paper and paperboard, more than six times the 1950 level. To visualize this, imagine paper piled so high it could reach the moon and back more than eight times.[69]

The forest situation is especially critical for the United States, where per capita consumption of wood and wood fiber is the highest in the world and growing. Americans consume more than 25 percent of the forest resources exploited on the planet, yet the nation is importing more and more forest products rather than growing them at home.[70]

A quarter of the wood cut in the United States is used in paper production, an amount that doesn't meet our pulp needs. And paper demand in the United States is swelling, with office consumption of paper rising about 20 percent per year. In 1996, more than 800 billion sheets of office paper were used in office copy machines around the United States—nearly 6,000 sheets of paper for every person in the labor force.[71]

Clearly the United States and other industrialized nations play a disproportionate role in the amount of paper consumed. In 1997, the average American used more than 700 pounds of paper and paperboard, double the rate of all industrialized nations and nearly seven times higher than the worldwide average. Developing nations use just 40 pounds of paper and paperboard per person, which experts say isn't enough to meet even basic education needs.[72]

Some industry analysts project a serious paper shortage before the decade is over, a shortage linked in part to population growth but largely to increased per capita use. A shortage in the paper supply will translate to rising prices, which will in turn dampen demand, encourage more recycling, and promote the substitution of other fiber sources.

Many Americans may be unaware that all over the world people use paper that is made from materials other than wood. Kenaf, for example, is a relative of okra and cotton that has been recognized for more than forty years as a high-quality fiber for paper production. Unless demand shifts to these and other materials, the planet will continue to lose forests and the services they provide, ranging from flood control and wildlife habitat to carbon dioxide absorption and oxygen production.

We fool only ourselves in believing we can replace what we've lost when we carve up the forests and bring down the ancient, old-growth trees. The "replacement forests" we plant in place of natural forests may be vastly different ecologically. Robert Paehlke, a professor at Trent University in Peterborough, Ontario, makes a larger point when he asks: "Should humankind remove and replace all the forests of the world? The issue for most environmentalists is not the economic value of forests, but whether all the world exists simply for our benefit."[73]

Wilderness and National Parks

Americans need only look to our prized national parks system to see the raw effects of population growth. The parks have experienced a one hundred–fold increase in the number of annual visits, rising from about 3 million visitors in 1930 to 30 million in 1950 to nearly 300 million in 2000.

At the same time, these places are subject to the pollution that rises from our major cities and urban areas. We've already discussed how air pollution dims visibility in our national parks. And whatever happened to freedom from noise? Snowmobiles roar through Yellowstone National Park in the winter, and sightseeing planes whir overhead at Yosemite. So many planes were flying over the Grand Canyon that environmental groups filed suit against the Federal Aviation Authority (FAA) to restrict the number of flights passing overhead. Back in 1987, when sightseeing flights down the canyon reached 50,000 a year, Congress had ordered the FAA and National Park Service to "substantially restore the natural quiet" of Grand Canyon National Park by creating no-fly zones and other restrictions on air tours. More than a decade later, however, the number of commercial flights over the canyon has doubled, and the park is even noisier. Park officials report that during daylight hours, a plane or helicopter can be heard every 90 seconds. Peregrine falcons have abandoned their nests, and bighorn sheep have been driven from their isolated mountaintops. The solution? The FAA capped the number of flights over the canyon at 100,000 a year. Some fix.[74]

I've worked hard to keep our wildest of lands wild, but more needs to be done. It's been shown that it can be done, and without too much pain. When I was governor in Wisconsin, my administration launched a stewardship program that conserves park and undisturbed lands, a program that continues to this day. Its initial funding source was a one-penny tax on a box of cigarettes.

I'm glad we have been able to preserve scenic mountain ridge views that are now vistas along the Appalachian Trail. As a cosponsor of the federal Wilderness Act, I am pleased that many other lands have been saved through that and other legislation. Among the accomplishments:

- Preservation of 104 million acres of wilderness.

- Growth of the National Forest System to 187 million acres, of which more than 32 million acres are protected as wilderness.

- Expansion of national wildlife refuges to include 92 million acres of land, triple the acreage in 1970.

- Designation of more than 150 wild and scenic rivers, with more than 10,000 protected miles.[75]

Wilderness provides Americans with much more than first-rate camping, hunting, fishing, and hiking. Wilderness cleans our air and water, provides important habitat for fish and wildlife, and serves as a natural laboratory where tomorrow's breakthrough medicines are discovered. As our population continues to grow, our wilderness

areas and parks will become increasingly precious, and efforts to preserve these places will be even more important. Designated wilderness areas today account for only 4 percent of the nation's total land
base and just 2 percent of the land in the lower forty-eight states. The
intent of the Wilderness Act was to protect such areas for all time,
yet wilderness areas across the nation are being challenged. The Bush
administration has gone head to head with environmentalists in seeking to drill for oil in the Arctic National Wildlife Refuge. Military
aircraft exercises interrupt the silence sought after by visitors and
wildlife at the Cabeza Prieta National Wildlife Refuge in Arizona.
Logging, drilling, and urban sprawl are other major threats.

In U.S. national forests, nearly 60 million acres of roadless wildlands remain, unfragmented areas that are fertile ground for wilderness designation but are likewise prized by the timber industry for
logging. This is land spread over thirty-nine states that should be preserved and should remain off-limits to road building and logging,
despite political pressure to the contrary. These lands include the last
great temperate rainforest in America, Alaska's Tongass National
Forest. We don't need the roadless areas for timber. Consider that only
4 percent of this country's timber comes from our national forests,
and only 6 percent of that—or one-quarter of 1 percent—would be
affected by conserving roadless areas.[76] Is it worth disturbing these
remaining wildlands to supply this small sliver of our nation's timber supply? Surely we cannot replace these rare natural treasures once
they are gone.

Meanwhile, grassroots understanding about the importance of
rainforests and our own old-growth forests has helped conservationists in their bid to preserve our national forests. In 1999 the Wilderness Society, in honor of the thirty-fifth anniversary of the Wilderness
Act, called for another 200 million acres of public lands to be protected in the National Wilderness Preservation System. The Sierra
Club in 1999 launched a campaign that called for the preservation
of much of the threatened wilderness areas along the route first taken
two hundred years ago by Lewis and Clark. In 1805, two years after
leaving the Mississippi River at St. Louis, the pair reached Oregon
and first glimpsed the Pacific Ocean. Some areas along their path have
already been preserved, such as the Little Missouri Badlands in North
Dakota and the Lolo Trail in Idaho and Montana.

The growth of the National Park System is certainly one of the biggest success stories of environmental progress during the past thirty years. Between 1960 and 1995, national parks grew from 26 million acres to more than 80 million acres.

But we need to take better care of these lands we have set aside for posterity. The soaring popularity of parks has created a new threat to their wildlife and preservation. Our fascination with exhaust-belching riding machines has carried over from our daily commutes to our weekend excursions. When we try to get away from it all these days, we take it all with us. This poses new threats to our parks and the roadless areas of our national forests. Off-road vehicles—dirt bikes, all-terrain vehicles, and snowmobiles—are overtaking our national parks and even trespassing on protected wilderness areas.

Yellowstone National Park in Wyoming is a vulnerable case in point. The park, notorious for being mobbed with cars in the summer, bans cars in winter. There are now more than about 82,000 visitors riding nearly 68,000 snowmobiles in the park each season, up from nearly 36,000 visitors riding more than 30,000 snowmobiles in winter 1973–74.[77]

A snowmobile's impact on the environment is staggering. Just one of these machines can emit as many hydrocarbons as 1,000 cars and as much carbon monoxide as 250 cars. Park employees are complaining of nausea, headaches, and throat irritations. Ted Williams wrote in *Audubon* that "fresh air has to be pumped into the entrance booths." Park officials now seek to limit snowmobiling within its borders and are backed by a grassroots group. As Yellowstone superintendent Michael Finley solemnly put it, "We are not debating some abstract scientific interpretation here. We are deciding if we are going to pass Yellowstone on to our children in good condition."[78]

As we think about the richness of the world in which we live—its forests, its clear blue waters, and all of its varied life forms—we must understand how our actions affect all of them.

We cannot be so arrogant as to think that we can do without any species on this fragile Earth of ours, or that our lives are not enriched by each of them. As respected naturalist Aldo Leopold wrote more than fifty years ago, "The last word in ignorance is the man who says of an animal or plant: 'What good is it?' If the land mechanism as a whole is good, then every part is good, whether we understand it or not."[79]

There is much work to be done. Despite all the progress that has been made, the words of caution I spoke back in 1965 are as urgent today as they were back then: "The great resources of America—the soil, the timber, the minerals, the wildlife—have sustained us for hundreds of years. But now we have to think about sustaining them. The frontier is gone. If we destroy these rivers and lakes, if we plunder these forests and rip up these mountainsides and foul this air and water, there will be no new green paradise awaiting us over the horizon. If we don't save the America we have today, I don't think we will have another chance."

5

An Invisible Threat

Since pesticides were developed in the 1940s, we have turned loose on the Earth a massive dose of compounds that can cripple or kill and which are tragically indiscriminate in their attacks.

—1970

Like most people, I didn't question it when the trucks pulled up and fogged the neighborhood with DDT (dichlorodiphenyltrichloroethane)— blanketing the plants, the trees, the grasses, and everything else. That was in 1960–61, some forty years ago, and they were spraying to wipe out mosquitoes in the community of Maple Bluff, located just outside Madison, Wisconsin.

We were living at the governor's residence at the time, and my sister, Janet Lee, was visiting. She put two and two together right away. "My God," she said. "This is crazy. Don't they realize this is a general medication of the whole community of life? They aren't just killing mosquitoes, they're killing birds, they're killing insects—they're killing all kinds of things."

I hadn't thought of it before, but she was right. They were spraying a highly toxic pesticide all over the neighborhood so folks could sit outside in mosquito territory and not get bitten. It seemed like a bad idea. But nobody back then knew how bad it ultimately would prove to be.

That incident prompted a proposal to ban the use of the pesticide DDT. I continued to push the issue when elected to the U.S. Senate, introducing the first proposal for a nationwide ban on DDT in 1963, though it took another ten years for that to happen. Several Wisconsin newspapers attacked the proposal, advising that "Sen. Nelson should leave science to the scientists." Well, you didn't have to be much of a scientist to understand that this was a dangerous, high-stakes game we were playing with nature.

Five years later, I was the first witness called to testify in a case that would make Wisconsin the first state in the nation to ban DDT. The Environmental Defense Fund had filed a petition asserting that DDT was a pollutant of Wisconsin waters. The trial opened December 2, 1968, with great fanfare under the dome of the state capitol in Madison. The crowd was so large that the first sessions had to be held in the Assembly chambers. Local newspapers, which had paid little attention to my earlier speeches against DDT, covered this event with banner headlines.

The case hinged on whether the plaintiffs could prove, under the state's water pollution law, that DDT met the definition of a pollutant of Wisconsin waters. After six months of testimony and a parade of twenty-five witnesses for the plaintiffs, state hearing examiner Maurice Van Susteren ruled it did. DDT was banned in Wisconsin.

When Rachel Carson's blockbuster book *Silent Spring* came out in 1962, it started a vigorous national dialogue on herbicides and pesticides that continues to this day. With her compelling prose, Carson dramatized for the American public the notion that saturating the landscape in DDT and other chemicals was harmful not only to plants and wildlife but to people as well. After her book came out, she came under heavy attack from scientists and representatives of the chemical industry who said she wasn't qualified to draw such dramatic conclusions.

At around the same time I ran into a distinguished scientist friend, Dr. Jim Crow, on campus at the University of Wisconsin–Madison and raised the point with him. "You know," I said, "all of these scientists are attacking her as being unqualified and so forth—what do you think of all this?"

"Well, she may not be qualified to reach such sweeping conclusions," he said. "And one can certainly question her qualifications." He paused then and added: "But in my gut, I know she's right."

By a wonderful coincidence, in July 2001—almost forty years after Dr. Crow put his trust in his gut—we met again on the university campus at a meeting of the Wisconsin Academy of Sciences, Arts and Letters. Fortunately, Dr. Jim Crow's gut was right, and DDT is gone.

You're killing bees. Now, if you kill the bees, who's going to do the pollinating? It's not a tough scientific question; it's just common sense.

When Carson wrote of major fish kills, and of robins, warblers, cardinals, and other backyard birds dying or unable to reproduce after exposure to DDT, she shed light on a growing body of research that was linking the pesticide to its insidious effects on wildlife. Based on the information that was emerging at that time, it was common sense to ban the chemical in the United States and throughout the industrialized world. The public knew it and ultimately made it happen. Since then scientists have tied DDT, as well as some other pesticides, to harmful effects in wildlife. And U.S. health organizations, such as the Environmental Protection Agency and Department of Health and Human Services, now classify DDT as either a possible or probable human cancer-causing agent.[1]

Common sense tells us these substances should be out of world circulation today. They are not. DDT continues to be produced in a handful of countries, including India and China. The pesticide is imported by more than two dozen other countries—primarily to fend off deadly malaria and other insect-borne diseases that afflict some of the world's poorer nations.[2]

Despite its notoriety, DDT is considered by some doctors to be the best answer to combating a lethal disease that claims many lives. In 1998, the World Health Organization estimated, for instance, between 300 and 500 million cases of malaria worldwide, causing about 1 million deaths per year.[3] DDT may be the treatment of choice in places where alternatives are not yet readily available, but there are many places where mosquitoes have become resistant to DDT—such as India and South America—and others where alternative measures have proved just as effective at reducing exposure.[4]

Why is DDT so destructive? Once introduced into the environment, the chemical casts a long, lingering shadow. Despite a federal ban on its use in the United States since 1972, DDT still leaves its mark thirty years later—long after Carson warned that a "chemical barrage has been hurled against the fabric of life."[5]

The pesticide at one time helped push our proud national symbol, the American bald eagle, to near extinction in forty of fifty states. DDT interfered with the bird's reproduction by weakening its eggshells to such a degree the shells were crushed under the weight of the nesting parent. After the banning of DDT and more than twenty years of federal protection, it has only been in the last few years that the number of eagles has risen enough for the government to upgrade its status from endangered to threatened.

Meanwhile, because of its long half-life, DDT residue continues to be found on fruits and vegetables. It also continues to accumulate in the fatty tissues of fish, birds, animals, and people, though these tissue concentrations have declined significantly in this country since the chemical was banned in the United States.[6] At the same time, the U.S. Fish and Wildlife Service reports that DDE (dichlorodiphenylethylene), a highly persistent breakdown remnant of DDT, is also taking its toll years later, accumulating in body fat and continuing to affect reproduction in bald eagle populations in places around the Great Lakes.

The federal government stopped the use of DDT, but, unfortunately, our reliance upon unsafe chemicals hasn't diminished. In fact, more than thirty years after the alarm was sounded about the hazards of saturating crops and communities with pesticides, harmful chemicals continue to be spread over farm fields and yards and around schools, hospitals, and parks as never before.

The EPA reports that approximately 20,000 pesticide products are currently registered for use in the United States, and about 890 active ingredients are registered as pesticides. Measuring active ingredients alone, the agency reports that in a typical year about 4.5 billion pounds of chemicals are used as pesticides in the United States. In 1997, for example, farmers, industries, government, and homeowners applied more than 4.6 billion pounds of pesticides in this country, the equivalent of 17 pounds per person.[7]

In short, EPA data show that pesticide application in this country has grown substantially since Carson's warning. And these chemicals are far from benign. An analysis of 1995 federal data by the Natural Resources Defense Council and the U.S. Public Interest Research Group found that the pesticides currently in use include acute toxins, suspected cancer-causing chemicals, and chemicals believed to disrupt the human hormone system. A 1993 National Academy of Sciences (NAS) study looking at the diets of infants and children noted that, "depending on the dose, some pesticides can cause a range of adverse effects in human health." Among these effects are cancer, injury to the nervous system, lung damage, and reproductive problems.[8]

In California, which alone is said to account for up to 25 percent of the nation's pesticide use, a stark picture of our chemical dependency emerges. From 1991 to 1998, more than 1.5 billion pounds of pesticides were applied in the Golden State, a state known for its environmental progressivism. Use of pesticides there more than doubled during that time, with no downward trend in sight.[9]

"Pesticide use trends show that California is hooked on toxic pesticides," reported Susan Kegley, a staff scientist at the Pesticide Action Network and lead author of the 2000 study *Hooked on Poison: Pesticide Use in California, 1991–1998.* "Use of the most toxic pesticides, including carcinogens, remains alarmingly high—indicating that the state is on the wrong track."[10]

Just as chemicals know no boundaries once released into the environment, pesticide application doesn't stop at our borders. At the same time that we're learning more about the biological hazards of pesticide use, chemical companies in the United States and other industrialized nations are increasing their export of pesticides forbidden for use within their own countries.[11] Worldwide, pesticide use has increased twenty-six-fold in the last fifty years, though fortunately this rate of growth is slowing. As insects and other pests build up resistance to pesticides, these compounds are increasing in strength, with their toxicity measuring anywhere from ten to one hundred times what it was in 1975.[12] Still, today's pesticides— though increasingly potent—thus far appear not to be wreaking as much havoc as DDT, a persistent organic pollutant banned by the United States and other developed nations as one of a "dirty dozen" that are considered the greatest chemical risks to humans, wildlife, and the environment.

Farming is far and away the greatest source of pesticide introduction into the environment. This has been true for a long time. Those working on a farm any time after World War II were aware that there were significant chemical residues on the food sent to market. But chemical companies told farmers not to worry, and doctors reassured their patients.

It was long after Earth Day that the general scientific and medical communities fully realized that children are a susceptible population outside as well as inside the womb. In fact, some pesticides may pose a greater risk to infants and children than to adults. Pound for pound of body weight, children drink more water, eat more food, and breathe more air than adults. Children in the first six months of life drink seven times as much water per pound as average American adults. Children ages one through five years eat at least three to four times as much food per pound of body weight as average American adults, and the air intake of a resting infant is twice that of an adult.[13] Indeed, it is only in the last decade we've learned that some foods contain nearly a toxic dose of pesticides if prepared for children. The potential damage done to recent generations by this early childhood exposure is, as yet, unknown.[14]

In no small part because of the diligence of environmentalists, these facts were recognized in the landmark federal Food Quality Protection Act, passed unanimously by Congress in 1996. The act revised and strengthened the nation's pesticide laws as well as set new standards for protecting infants and children. Specifically, Congress called on the EPA to reassess more than 500 pesticide chemicals that are registered for use on food crops and determine their safety based on a new scientific standard of "reasonable certainty of no harm." Once the EPA has determined the safety of these chemicals, the law requires the agency to reassess and adjust the legal limits for some 9,600 pesticide tolerances—or limits—in foods, giving particular consideration to exposure risks for children.

In passing the law, Congress adopted key recommendations from the 1993 NAS report on pesticides in the diets of infants and children. In what may seem an obvious point, the academy stressed that children are not "little adults," and that therefore their small, developing bodies metabolize and handle chemicals differently than do adults. The effects can be serious. "Compared to late-in-life exposures, exposures to pesticides early in life can lead to a greater risk of chronic effects that are expressed only after long latency periods have elapsed. Such effects include cancer, neurodevelopmental impairment, and immune dysfunction," the academy wrote. The NAS called for better information on infants' and children's dietary exposure to pesticide residues and the potential effects, and it recommended establishing pesticide tolerance levels to safeguard the health of infants and children.[15]

Passage of the Food Quality Protection Act is one of the most notable environmental gains of the 1990s; it serves as a major step toward getting a better handle on the effects of the chemicals we spread throughout the environment. As a result of the act, in 1999 the EPA banned or restricted two of the oldest common chemical compounds in use as pesticides: methyl parathion and azinphos methyl. Methyl parathion, an insecticide used by fruit and vegetable growers that had remained on the market for years, has been characterized by the Consumers Union as the "riskiest single pesticide detected in the U.S. food supply."[16] The EPA itself acknowledged the chemical as "one of the more potent organophosphates" in announcing its phaseout.[17]

The EPA also phased out home and garden use of the insecticide chlorpyrifos. Perhaps best known by the trade name Dursban, the pesticide is used in 20 million U.S. households each year—incorporated into hundreds of everyday household chemicals ranging from flea collars to indoor bug sprays to lawn and garden chemicals. The EPA determined after a lengthy review that the chemical poses a risk to children because of its potential effects on the nervous system and possibly brain development.[18]

As the EPA moves forward with its new charge from Congress, however, some warn it is not moving fast enough to protect chidrens' health.[19] In 1999, and again in 2000, the Consumers Union reported that pesticide residues on some of the foods children eat every day—such as apples, peaches, pears, and spinach—often exceed safe levels, even as they tested within established U.S. legal limits for pesticides on those foods. "Legal limits do not define safety, and residues of some chemicals on some foods would frequently expose a young child to a dose greater than the U.S. government's official estimate of the 'safe' daily intake of those pesticides," the report concluded.[20]

Chemicals that are released routinely into the environment course their way to our rivers and water supplies. Pesticides leach into underground aquifers to contaminate drinking water sources and find their way to creeks, rivers, and lakes, where they can seriously harm or kill aquatic life.

Yet pesticides are only a small part of a much larger chemical assault on the environment that began in earnest after World War II. The last half-century has seen large-scale production, use, and discharge of synthetic chemicals into the environment. These chemicals are everywhere, with more than 80,000 produced in the United States alone.[21] Data from the EPA's annual Toxics Release Inventory show that in 1999, for example, industry released nearly 8 billion pounds of toxic chemicals into the nation's air, land, and water.[22]

Acid rain, a major threat in the early 1980s because it can kill fish and contaminate water, has been curbed, yet it remains a problem. Many areas saw improvement in the 1990s thanks to EPA-mandated cuts in industrial sulfur dioxide emissions from coal-burning power plants and in nitrogen oxide emissions from vehicle tailpipes. Still, our lakes and mountain forests have seen little benefit.

Nationwide, more than two thousand advisories warn anglers to limit or avoid eating fish from hundreds of inland lakes stretching from Maine to Minnesota to Florida. Mercury, emitted in the greatest quantities by coal-burning utilities, is the greatest source of contamination in these lakes. Also playing a role are PCBs, chlordane, dioxins, and DDT or DDE.[23] Fish advisories, however, fail to protect wild animals high on the food chain that feed on contaminated fish.

Joe Heller/*Green Bay Press-Gazette.*

In short, there doesn't appear to be any place on Earth, no matter how remote, left untouched by toxic chemicals. Earth's last holdout, the Arctic north, has been sullied even in the years since Carson penned the following passage in *Silent Spring:* "To find a diet free from DDT and related chemicals, it seems one must go to a remote and primitive land, still lacking the amenities of civilization. Such a land appears to exist, at least marginally, on the far Arctic shores of Alaska. When scientists investigated the native diet of the Eskimos in this region it was found to be free from insecticides."[24]

In recent years, Carson would be distressed to find that the planet's polar regions are repositories for persistent chemicals, despite being located more than a thousand miles away from major industrial centers. Because of a diet high in fatty foods such as whale, seal, and polar bear meat, Inuit women in the remote regions of the Arctic have been found to carry higher concentrations of PCBs in their breast milk than other Canadian and U.S. women.

The knowledge we have gained since the alarm was first sounded about these invisible contaminants is unsettling and, like a good college education, shows us how little we really know.

We may not be able to see this formidable adversary, but we can see its effects. Young gulls, herons, and terns look out over crossed bills, backward wings, and feet that grow sideways or not at all. Birds ignore their broods or nest with others of the same sex. Some eggs never hatch, while well-fed chicks mysteriously waste away. Beneath the water's surface, fish eggs and fry perish. Salmon are found with enlarged thyroids, and walleyes are burdened with tumors and cancerous lesions. These are just some of the disturbing abnormalities found today in areas of the Great Lakes. This largely self-contained and slow-moving lake system has long served as an experimental laboratory of sorts for scientists. The lakes act as a giant holding pen for a variety of persistent toxic chemicals that have been discharged or carried by air, then stored in the lakes or their tributaries—often for decades.

There is further evidence of chemical contamination from coast to coast and around the globe: penis abnormalities and population declines in alligators of Florida's Lake Apopka, which were heavily exposed to DDT; diseased and deformed mink in the United States

Trace chemical contaminants can produce physical deformities and behavioral changes in animals such as this Great Lakes cormorant. Photograph courtesy of *Green Bay Press-Gazette, and Tom Erdman, Richter Natural History Museum.*

and river otters and marine mammals in Europe. All of these defects are linked to environmental contaminants.

A leading, if controversial, theory is that these abnormalities in wildlife are the result of chemicals that do their damage by disrupting hormones, chemical messengers in the body that direct the reproductive, nervous, and immune systems. This was an unforeseen hazard back when the first concerns about chemicals were publicized in the 1960s, but these so-called hormone disrupters have been the subject of intense research since the mid-1990s.

This research has shown that chemicals in the environment such as PCBs and dioxins can disrupt the reproductive, nervous, and immune systems of animals in the wild and in the lab. Some scientists suspect that the body perceives these manmade chemicals as hormones—such as estrogen—and responds as it would to its own natural hormones.

Theo Colborn, the World Wildlife Fund senior scientist whose 1996 best-seller *Our Stolen Future* first theorized about chemicals in the environment masquerading as hormones, said she became interested in the effects of chemical contaminants on wildlife while working toward her Ph.D. at the University of Wisconsin–Madison in the early 1980s. Reflecting briefly on environmental changes since 1970, Colborn noted that although releases of some chemical contaminants have gone down, it hasn't been enough to offset all of the problems. "In the 1970s people discovered that the Great Lakes wildlife were in trouble, and even in the late 1990s the same species are still in trouble," she said.[25]

The fact that deformities and reproductive problems are showing up in wildlife high on the food chain, such as in fish-eating birds and otters, has Colborn and other scientists concerned that these same chemicals also may threaten the folks at the top of that chain: people. Because hormones control the body's reproductive system, scientists are exploring whether these environmental chemicals might be to blame for some human reproductive problems, such as falling sperm counts, infertility, and testicular, breast, and prostate cancers.

Most importantly, however, because hormones control development from conception to birth, there is concern about the extent to which PCBs and other hormone disrupters interfere with the development of children in the womb and even after birth throughout breast-feeding.

We won't know the potential dangers of chemicals until we change how we go about testing them, according to Colborn. She notes that chemical testing in the United States has traditionally focused on cancer rather than noncancerous effects or the potential for those effects to be carried from one generation to the next. Furthermore, though synthetic chemicals are everywhere we look, not one has been tested for hormone-disrupting effects fully based on the knowledge we have today.

"We've tested some of them—but not very many—for cancer. We have looked for obvious birth defects . . . and mainly to see if the rats could reproduce," Colborn said. "But we never bothered to look to see what happened to those pups when they grew up as a result of prenatal exposure. We have never tested a chemical from the moment the sperm enters the egg until an individual is born or

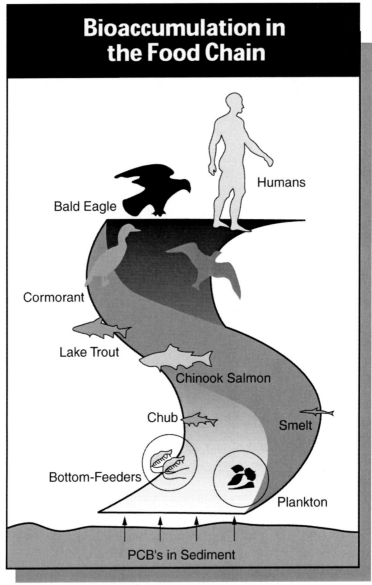

Chemical contaminants accumulate.
Of all creatures in the food web, humans are among those most subject to high levels of
contamination by certain types of chemicals such as PCBs. This is a result of bioaccu-
mulation, a process in which each level of the food web attains higher concentrations.
Source: U.S. Environmental Protection Agency.

hatched, and we don't have one screen or assay today to test the chemicals for those effects. And until we get some screens and assays to do that, we risk adding more hormone disrupters to the environment. And industry's going to tell us that it doesn't matter because they're out there in very, very low doses."[26]

Colborn's provocative book, and its ability to heighten public interest and awareness in the effects of chemical pollution in the environment, prompted federal funding to expand research on potentially hormone-disrupting chemicals. In 1998, Congress called on the U.S. Environmental Protection Agency to look for hormonal effects in 100,000 chemicals used in common household products.

Congress, in conjunction with the EPA and the Department of the Interior, also called on the NAS to convene a panel of some of the nation's leading experts on the subject to look for answers. The sixteen-member scientific panel concluded in 1999 that higher levels of persistent, hormonally active chemicals in the environment such as PCBs are associated with various troubles afflicting fish and wildlife, including reproductive problems, aberrant behavior, deformities, and disappearing populations. But the panel stated that the way in which these chemicals do their damage remains unclear, leaving the issue of whether they interact with and trick the hormone system an open question for now. The panel also concluded that there haven't been enough studies conducted to link hormonally active agents in the environment with the falling human sperm counts reported in some parts of the world, or for the increasing incidence of human breast and prostate cancers.

It did, however, cite research by two Michigan scientists looking at the effects of PCBs on babies as the most comprehensive human studies of their kind to date. Psychologists Joseph and Sandra Jacobson of Wayne State University found poor short-term memory in infants and young children whose mothers had elevated PCB levels in their umbilical cord blood. The couple published a follow-up study in 1996 of the same children eleven years later, reporting that higher PCB exposure in the mothers corresponded with lower IQ scores, poor memories and attention spans, and below-average math and reading skills in their children. The NAS panel found that those and other studies involving humans and animals "indicate prenatal exposure to PCBs can cause lower birth weight and shorter gestation, and has been correlated with IQ and memory deficits as well as delayed neuro-muscular development."[27]

The panel's general conclusion about the interactions between low-level chemical contaminants in the environment and people underscores the relative infancy of this critical area of research. After an exhaustive review of the scientific literature, the panel determined more research is needed.

Colborn said this research not only has a long way to go but also requires a whole new way of looking for and assessing chemicals' effects on people and the environment. "What's coming out of the academic laboratories is an entirely new kind of toxicology," she said. "It is looking at reproductive success two and three generations down the line. It is looking at behavioral and learning impairment in these offspring, as well as their ability to respond to immune system challenges. It is looking at those systems that make us human—and it appears that at very, very low doses these chemicals interfere with those processes."[28]

What we do know is cause enough for concern. While the scientific debate rages on, these chemicals persist in the environment and continue to be released into the air and water. In some developing countries, PCBs are still in use—and in Russia, they are still being manufactured. Dioxins are also present throughout the environment. At the time of the Vietnam War, I introduced legislation to ban Agent Orange when it was learned that the defoliant used by the U.S. military contained a highly toxic form of dioxin known as TCDD. We were spraying millions of gallons of the herbicide to defoliate trees and shrubs along rivers and in forests so the enemy couldn't hide. It was a version of environmental warfare. Agent Orange was banned, but dioxins continue to be released in the United States and around the world as unintended by-products of incineration, paper bleaching, and other processes.[29]

The persistence of these and other chemicals means they linger in the environment for decades. And once there, they don't stay put. As the Arctic's remote Inuit population shows, PCBs and other persistent chemicals travel far and wide, riding air currents, moving through water, and traveling up the food chain. These persistent organic pollutants, known as POPs, also collect in the sediment of hundreds of miles of the nation's waterways. This is true of the Fox River in Wisconsin, where regulators say it will be a century before contaminant levels drop low enough on their own for all fish to be eaten safely. There, cleanup plans are still evolving after more than a decade of delay by paper mills that discharged PCBs into the river through papermaking and recycling processes before the chemicals were banned in the late 1970s. Starting in "Paper Valley"—home to the world's largest concentration of paper mills—the contamination runs along 40 miles of the river before spreading into the bay of Green Bay and beyond to Lake Michigan. A similar case occurs along 200 miles of the historic Hudson River in New York. In 1998, EPA administrator Carol Browner stressed the importance of removing these contaminants from the nation's waterways as she addressed members of the New York State Assembly, where cleanup of the Hudson was stalled through protracted foot-dragging by General Electric, the source of the river's PCBs. "We do not have every single answer, nor every single piece of data," she said. "But clearly, the science has spoken: PCBs are a serious threat—a threat to our health, a threat to our environment, a threat to our future."[30]

In 1970, I wrote the following about the scourge of pesticides, based on available scientific research:

The already massive and still accumulating evidence on pesticides makes it clear that these toxic compounds have become one of the most serious problems of our environment and are threatening even greater worldwide damage. Pesticides have concentrated to the far ends of the Earth. They're killing fish and wildlife. They have inhibited fish and wildlife reproduction. High pesticide residues have pushed some fish-feeding birds and other animals to the edge of extinction; and now, there is increasing concern and evidence about the threats posed to man.

The evidence we have collected since then has shown these threats to be even graver than we thought. We know that once let loose on the Earth, these substances don't just disappear. In less than a century's time we have set up a vast scientific experiment on this planet, casting ourselves and all other living creatures as involuntary test subjects.

As was true thirty years ago, we don't know where this might lead. We do know, however, that the problems have not eased with the passing of time. Although we ceased producing DDT and PCBs decades ago in this country, these and other synthetic chemicals remain with us still. The great water birds with their crossed bills and twisted legs tell us so, as do their wasting chicks. Beneath the water's surface, the fish whisper the same refrain. We, ourselves, have not escaped unscathed, though the extent of our injury is not yet apparent. Like unbidden guests, these chemicals inhabit our bodies and those of our children—even before birth, we now know.

We can wait, as some suggest, until we have conclusive evidence of the damage these substances do when they take up residence within us. Or, we can take our cue from the birds and animals that already are experiencing higher exposures and greater sensitivity.

The public has shown itself able to comprehend the potential that awaits us in this evolving world laboratory. It has forced lawmakers to respond with chemical bans, greater scrutiny, and improved accountability for what we spread throughout our environment. More recently, the public has begun to grasp the subtlety of what these substances might mean to future generations as each passes on a higher dose of chemicals to the generation that follows.

We must do more, however. We are still a society that is heavily reliant upon chemicals, yet we are largely ignorant of their long-term implications. We must continue to find alternatives. We must learn more about the latent hazards found within the chemicals we spread. We must be cautious. Above all, we must not blind ourselves to nature's recurring warning.

PART 3
Environmentalism:
Then and Now

6

Complacent Planet?

*The battle to restore a proper relationship between man and his envi-
ronment, between man and other living creatures, will require a
long, sustained, political, moral, ethical and financial commitment
far beyond any effort made before.*

—1970

A long, sustained commitment to the health of the environment is
more urgent today than it was in 1970.

Yet, after all of the early success of the environmental movement
in the 1970s and its challenges during the 1980s and 1990s, where
is it today? At a time when environmental problems loom larger than
ever, necessitating an even broader commitment to activism and stewardship, is the commitment still there?

Many qualified academics have studied the issue, pollsters have polled
about it, and any number of folks have theorized about it. This is a good
time for all of us to step back and take the pulse of the environmental
movement. But doing so is as much a gut check of our own attitudes
and practices as it is an assessment of a mass movement. Indeed, some
of the worst forms of environmental degradation continue to be the cul-
mination of smaller, individual insults rather than a handful of headline-
grabbing disasters. In short, caring for the Earth begins at home with each
of us long before it can ascend to a national and global attitude. How
far along we are in our individual stewardship and whether we can go
the distance is a question each of us must ask ourselves.

"I Am an Environmentalist"

The stage is set. The public is primed for action, but it does not
yet feel compelled to act.

Strictly by the numbers, the environmental sentiment in this country is as strong as ever. National polls consistently show strong support for the environment, with more than half of those surveyed telling pollsters they are "sympathetic to environmental causes, but not active," and 15 percent to 20 percent characterizing themselves as "active participants."[1]

Americans are well enough aware about environmental problems that when asked how strongly they agree or disagree with the goals of the environmental movement, 83 percent say they agree, and only 15 percent disagree. Notably, the movement trails only the civil rights and women's rights movements in levels of agreement.[2]

What about claims that environmentalists are extremists who exaggerate environmental problems? Ask the average person if his or her fellow Americans are "too worried" about the environment, and only 10 percent agree. Almost 60 percent say the American public is "not worried enough."[3] About the same percentage believe the U.S. environment is getting worse, rather than staying the same or getting better.[4]

We all know the economy is important, and we all have bills to pay. Yet, as has been true for at least two decades, 57 percent of Americans said in 2001 that they would favor protecting the environment even at the risk of curbing economic growth. This commitment to environmental protection through good and bad economic times has held true during the more than ten years the Gallup poll has been asking the question. Older survey series trace it back to the 1970s.

The American public's support for environmentalism is evidenced in a number of ways. Memberships in the national environmental organizations—such as the Audubon Society, Nature Conservancy, Sierra Club, World Wildlife Fund, and The Wilderness Society—have grown significantly since the first Earth Day.[5] More importantly, at the state and local levels today, hundreds of grassroots environmental organizations have sprouted and flourished in all parts of the United States. They are the local guardians of nature's works and represent the real vigor and political clout of the environmental movement.

With so much public support for the environment, why aren't there greater public pressure and progress toward preserving open space, natural habitat, and wetlands and toward curbing runoff pollution from farms and urban streets? Why are SUVs and pickup

"THEY USED TO BE THOUGHT OF AS CRACKPOTS."

Printed with the permission of Sidney Harris.

trucks increasingly the vehicles of choice among Americans, despite their well-known fuel inefficiency? Why does the United States emit more greenhouse gases per person than any other industrialized country in the world? And why does Americans' appetite for material goods, already the most voracious in the world, continue to rise?

The problem is keeping environmental issues not only visible but also a higher priority. People tend to view the environment as an important issue to monitor, but not urgent enough to require immediate attention. Politically, the attitude is that environmental problems can be put aside temporarily—while problems such as unemployment, crime on the streets, and the economy must be addressed immediately.

A lack of strong presidential and congressional leadership is a major part of the problem. If the president and Congress are not talking about meeting the challenge of sustainability, the public cannot be blamed for failing to heed those who warn of its urgency. At the same time, numerous books and articles have been written that state or imply that the major environmental issues have been solved, and these publications have received widespread, positive publicity. It is natural that the public, the media, and others want to believe the good news, as do special-interest groups.

The polls show that although Americans generally support a healthy planet and say more should be done to protect it, they clearly draw lines about how much they will spend and what habits they will change in order to help the planet. In this regard, the public has been lukewarm about the environment, with a noticeable low point in the mid-1990s.

The polling firm Roper Starch Worldwide has been tracking the "green" buying habits of Americans since the first mass marketing of such products in 1990. In 1997 it reported: "The most environmentally dedicated group hasn't declined, but the group most willing to put its money where its mouth is has shrunk dramatically. The most apathetic group has grown." Specifically, the report showed a drop in consumers' willingness to shell out up to 20 percent more for green products, dropping from 11 percent of adults in 1990 to 5 percent in 1996.

A follow-up study in 1999 found that those consumers willing to spend more for environmentally friendly products were even more measured in that willingness: paying on average just 8 percent more for energy-efficient major appliances, 6 percent more for electricity-saving home computers, and 6 percent more for biodegradable plastic packaging, recycled-paper products, and cars that are a third less polluting.[6] In short, the consequences are too vague and the calculation is too difficult for the typical consumer to spend much more for

green products. Ultimately, factors such as price, quality, and convenience are winning out at the checkout counter.[7]

A decision by Home Depot in 1999 to stop selling wood products from environmentally sensitive areas by the end of 2002 shows that consumers can and do mobilize against certain environmentally harmful practices, however. The company's move was, in part, a response to Sierra Club members mailing 20,000 postcards to Home Depot's CEO, along with pressure from Native Americans and other groups.[8] Another sign of hope is the "green building" movement, a trend toward using ecologically sustainable materials in residential and commercial buildings, driven in part by health concerns regarding chemicals in home interiors.

Meanwhile, Americans' unwillingness to sacrifice when it comes to their beloved automobiles has been proven again and again when it comes to changes and uncertainty in gas prices. The higher gas prices in early 2000—resulting from a decision by the oil-producing nations to cut the oil supply—prompted an angry outcry from American consumers. Always loathe to spend more for gas, though still paying less at the pump than people in other countries, many Americans were hit especially hard by the increase because they had invested in popular but gas-guzzling pickup trucks and SUVs. Similar to their reluctance to spend more for energy-efficient appliances or to opt for more fuel-efficient cars, many Americans regard the environment as a less worthy, "fuzzy," or expendable cause. It is worthwhile exploring why this is the case.

"Low-Salience Issue"

More than thirty years after the environmental movement hit full stride, many today believe the urgency that propelled that movement is gone—even as support remains high. The environment is considered a "low-salience issue" according to those who conduct opinion polls. Indeed, in poll after poll the environment is rated "very important" by two-thirds or more of those surveyed. But when compared with other national concerns such as education, health care, Social Security, and the short-term condition of the economy, the environment consistently is pushed out of the emergency room and into the waiting room.

In early 2001, Gallup reported that 26 percent of those surveyed said it was "extremely important" for the president and Congress to deal with the environment that year. And the environment was the public's top concern with respect to the future, looking ahead to the next twenty-five years. Yet when ranked with other problems, the generic issue of "environment" ranked sixteenth on Americans' list of the most pressing problems facing the country today—well below education, crime, and health care.[9]

Another early 2001 poll, this one conducted jointly by CBS News and the *New York Times,* reported that just 5 percent of respondents ranked the environment as the "most important" problem for the government to address that year. Energy concerns spawned by that year's California blackouts, a clear natural resource issue, scored high among problems confronting the nation. Still, when cast against a backdrop of other challenges the country is facing, environment ranked seventh on the poll's list of twelve national problems.

Why this complacency in the face of today's profound ecological challenges? Doubtless, there are a variety of forces at work, some of which are contradictory. One key factor is the environmental movement's success at addressing much of the visible pollution in this country since the 1960s and 1970s. In those years, the public was educated dramatically by oil-covered beaches, greasy multicolored streams and lakes, brown clouds of smog, and littered roadsides. People saw the huge challenges of air pollution and water pollution and tackled them, making significant headway.

Lending credence to the idea that people tend to care most about what they experience directly is the fact that these sorts of problems continue to be the ones that worry Americans the most in polls. For example, pollution of drinking water, rivers, and lakes, air pollution, and toxic waste contamination consistently rank among the public's greatest environmental concerns.[10]

Americans also are influenced by what they see in the media, and the environment was a hot story in the 1970s. Environmental issues not only enjoyed strong public support but also presented a new cause and an easy story to tell. Some reporters and photographers of that time say covering the environment could be as simple as heading down to the local river, sticking a hand in the water, and photographing it as it emerged covered with muck.

When these most graphic forms of pollution were cleared away, however, so too was much of the local media attention that thrived on compelling pictures of trashed beaches, brackish water, and stained skies. PCBs, fine particulates, and global warming don't make good pictures—if they can be photographed at all—and the general public usually doesn't see the direct, immediate effects of these problems when they walk out the front door.

Most environmental issues today are more complex and more difficult to communicate than they were back then, for both the media and the public. Today we have carbon dioxide buildup contributing to global climate change, invisible chemicals that are harming random members of future generations, minuscule air pollutants that aggravate our lungs, and species we never knew existed disappearing off the face of the planet. All of these subjects require time to explain and more scientific literacy than was needed to understand shores full of dead fish.

These factors, coupled with the fact that many of the most pressing environmental issues are slow burners that will go on for decades—as opposed to the breaking stories that are here today, gone tomorrow—help explain why media coverage often doesn't convey the depth or urgency of environmental problems.

Yes, we can take heart in the fact that environmental journalism exists today throughout all branches of the media. I knew of just one environmental reporter on a daily newspaper before 1970, Gladwin Hill, who wrote for the *New York Times*. Today, most major daily newspapers have a designated reporter who covers environmental issues. Still, many journalists report a long-term trend of declining interest among editors in the daily fare of environmental reporting. Some reporters say they lack adequate time or space to research and report the issues, while others complain that when they do have sufficient time, their stories are relegated to the back pages of the newspaper or late in the broadcast.

A survey of 496 environmental reporters in 1996 by Michigan State University's Environmental Journalism Program found editors often assign these reporters to cover non-environmental stories that are considered more newsworthy. Surprisingly, the vast majority of environmental reporters responding to the survey said they spent only half of their time covering environmental issues.

Toss in the conflicting evidence confronting the public on so many environmental stories, and it becomes clearer why the public does not yet feel compelled to act on behalf of the environment. The complexity of today's environmental problems, combined with the relative infancy of some of the research in these areas, has set off rancorous debates within the scientific community that tend to erode public confidence.

The arrival of such issues as hormone disruption and global climate change has taken the American public into the laboratory and exposed them, many for the first time, to the scientific process. This in itself is confusing, because of the scientist's practice of speaking in shades of gray uncertainty rather than the black-and-white certainty the public so often demands. Absolute proof will forever elude scientists in the complex experiments being conducted on the natural world, however. Scientific opinions and consensus are based on the overwhelming evidence of the time, and the scientific process will continue to be one of second-guessing and revising today's results and theories. To an uninitiated public so conditioned to the definitive pronouncements of politicians and marketers, this internal scientific debate smacks of an indecision that can be easily dismissed or ignored.

"Problem Solved"

The progress since the launching of the modern environmental movement may have had another unforeseen consequence: creating the impression that most of the nation's environmental ills have been dealt with already.

Wrong as that is, it's not hard to understand how that sentiment could develop. Environmentalism gave rise to any number of watchdogs whose principal mission has been to guard us from environmental harm. In 1970 the U.S. Environmental Protection Agency, the biggest watchdog of them all where the environment is concerned, was formed to enforce federal laws and regulate activities that threaten the environment and public health. State and local environmental agencies also were formed, and the few in existence were greatly expanded.

As mentioned earlier, the 1970s also saw the creation of more than two dozen federal environmental laws designed to further the cleanup of our air and water, protect endangered species, register pesticides, and protect our drinking water, among other things. At around the same time, hundreds of grassroots environmental groups sprang up to police their local communities, and older hunting, fishing, and birding groups increased efforts to protect habitat. New national environmental groups were created, and many long-standing groups saw their memberships grow.

Given all this attention to the environment, average Americans might find it hard to believe that most chemicals in pesticides have not yet been fully tested in accordance with federal requirements made in the 1970s. They may be surprised to learn that regulators have approved bioengineered corn and soy but are uncertain how these crops might effect the ecosystem, or that many a town board fails to give serious consideration to environmental impacts when it approves a sprawling new subdivision.

Indeed, at the start of the new millennium more than half of Americans said they were generally satisfied with the state of environmental protection in the United States, and two-thirds of Americans said we have made at least minor progress since the first Earth Day.[11] An earlier Gallup poll, from 1999, also found a public generally at ease with environmental protection—a mood that characterized the country for much of the 1990s. "Americans have grown increasingly satisfied this decade with the nation's environmental protection efforts," Gallup reported. "A majority still sees room for improvement, but there is a growing perception that there has been progress in dealing with environmental problems and that society—particularly government and the public—expresses a sufficient amount of concern for the issue."

This attitude has been borne out in consumer behavior, as was reported by *American Demographics* in a story about the 1997 Roper Starch Green Gauge report: "Fewer people use biodegradable soaps, avoid aerosols, read labels for environmental impacts, buy products made of or packaged in recycled materials, buy refillable packages, (and) avoid buying from ecologically irresponsible companies. . . . In one way, this seeming apathy is a backhanded compliment. It's as if many consumers take for granted that today's mainstream products treat the planet kindly. They feel they can relax their vigilance."[12]

The sense of security that comes from knowing government agencies and activists are watching over the environment and a poor understanding of what causes some of our biggest environmental problems have combined to create the impression among many that nature doesn't need our continued vigilance. This phenomenon, juxtaposed with today's staggering environmental challenges, may well have tragic consequences.

Failing Grade

Our awareness and understanding of environmental issues play a critical role in our level of concern. With that in mind, I have long been a proponent of educating people, from kindergarten on up, about nature and its interconnection with our lives and those of other creatures on the planet.

The 1995 Merck Family Study by the Harwood Group, which looked at Americans' attitudes about consumption, values, and the environment, underscored the importance of understanding this connection. "People perceive a connection between the amount we buy and consume and their concerns about environmental damage, but their understanding of the link is somewhat vague and general," the researchers reported. "People have not thought deeply about the ecological implications of their own lifestyles; yet there is an intuitive sense that our propensity for 'more, more, more' is unsustainable."[13]

Public knowledge about the environment is far beyond where it was several decades ago, as most students begin learning about the environment in kindergarten and continue learning through their college years. A true understanding of our environment, however, requires a continuing education if we are to keep pace with its changes and scientists' evolving knowledge.

Achieving this level of understanding requires demonstrating how the environment is an integral part of our lives and those of other creatures, not reducing the subject to mindless recitations of facts and statistics. This shouldn't be hard to accomplish, with nature's classroom beckoning just outside the door. Yet sometimes this point is missed.

I think of my son, Happy, coming back from grade school years ago. He had a book and twenty-five questions on a mimeographed sheet. He answered most of the questions but was baffled by two or three others. "How long is the Orinoco River?" one of the questions asked. "Who cares?" I thought. I found the answer in the book and wrote it in, and that's the first and last time I've heard a question about it. The teacher was turning off the children's desire to learn by asking obscure, boring, unimportant questions.

People need to understand general principles, not memorize statistics. If you teach people, as John Muir wrote, that everything is connected to everything else, they learn the principle involved. That understanding will guide their choices and actions throughout life much more than will any amount of memorization. The connection is there—we need only point it out.

Environmentalists

Despite the broad-based public support for environmentalism, a number of Americans—the press included—have been critical of the environmental movement for not attracting a broader, more diverse support base. This criticism is not well founded.

It is a fact, as is frequently asserted, that most environmental activists are white, well educated, and affluent. It is also a fact that for most minorities and their leaders the most urgent priorities involve issues of discrimination, jobs, equal justice, equal opportunity, and equal treatment in all aspects of their lives. These are pressing priorities, and there are limits to the amount of time and resources available to address them all. But that doesn't mean minorities or the poor are any less concerned or less sensitive to the impact of pollution. Indeed, it is the laborer living in the shadow of industry smokestacks, the black man living near an inner city truck route, and the Latino worker exposed to pesticides in the field who feel the sharpest bite of our disposable society. Polling data show that concern about the environment is no less among minorities or among the poor. A 1999 Gallup poll, for example, reported comparable results for whites and nonwhites, for the highly educated and the less educated. The fact is that there is almost universal concern for the state of the environment, and every environmental organization I know of welcomes members and support from every cross-section of society.

Reaching out is clearly on the minds of those at the forefront of the environmental movement today. At a conference in 1999, international wilderness expert Vance Martin urged the wilderness movement to heed America's changing demographics. Speaking at the Wilderness Horizons Conference in Ashland, Wisconsin, Martin noted that by 2010, the majority of Americans ages twenty-five years and younger will be nonwhite, and by 2050, the majority of the entire U.S. population will be nonwhite.[14]

The environmental movement must have the broadest base possible if it is to succeed in protecting the natural world. Maintaining a clean environment benefits rich and poor, black and white. No group has a monopoly interest over clean air or water. Likewise, everyone suffers equally if a drinking water source is contaminated, a swimming beach is closed, or the air is sullied.

That said, it is the polluted inner cities and gritty industrial neighborhoods that have been most effective in attracting minority involvement to the environmental cause—propelling the environmental justice movement in the 1990s. The leaders of this movement are influencing public policy, industry practices, and private foundation funding, among other things. Detroiters Working for Environmental Justice, for example, helped close a medical waste incinerator in a predominantly black neighborhood of the city, where children were hospitalized for asthma at a rate that was among the state's highest. On other fronts, a half-dozen environmental justice centers and legal clinics have been established around the country, and environmental justice courses and curricula can be found at nearly every university in the country.[15]

Just as decaying cities and urban environments prompted minorities to rally for environmental justice in the 1990s, the declining state of the global environment caused the religious community to join hands with environmentalists during the last decade. Although the two groups were strange bedfellows for years, the union was borne of a realization by scientists and environmentalists that the religious community could be a powerful ally in nurturing the public's environmental conscience. At the same time, the alliance has helped nurture the spiritual side of environmentalism—a side quite different from the one that looks to scientific fact as the primary impetus for conservation. The common ground is Earth: viewed as either fragile ecosystem or God's creation, it is one and the same world when threatened.

Paul Gorman, executive director of the New York–based National Religious Partnership for the Environment, said it is a natural alliance. "When people really take this to heart and soul, they realize how profoundly a religious concern it is. It's every moment under the stars. It's every moment that you've sat by a stream and felt spiritual renewal."[16]

Environmental groups, unfortunately, don't have the means to recruit members of any interest group exclusively. Instead, they cast a broad net, hoping to reach the widest audience possible with a universal message meant to resonate with all people sharing a common interest in the world in which they live. Scott Darling, of the Vermont Department of Fish and Wildlife, had the right idea when he once said that in the "big tent" of conservation, there's room enough for everyone: "It's a very big tent," he said. "Does it include every person in the world? I don't think so, but ideally it will in the end."[17]

The "Brownlash"

For every action there is an equal and opposite reaction. This principle of physics could be used to characterize the environmental movement during the 1970s and 1980s. If the 1970s saw the action, the 1980s saw the reaction. And a strong reaction it was. Business and industry saw themselves as victims of the sweeping environmental reforms of the 1970s and launched a powerful counteroffensive the following decade. Property rights and "Wise Use" groups (two well-funded and conservative countermovements that came of age during this 1980s backlash) and other anti-environmental groups also gathered force, helping to challenge environmental regulations at every level. Stanford ecologist Paul Ehrlich termed the backlash a "brownlash," a reference to the tangible effects of this assault on our environmental laws.

Corporate America had an ally in President Ronald Reagan, a longtime corporate spokesman for General Electric who worried about the effects of environmental regulation on the national economy. Many of the new environmental laws were re-evaluated from a superficial economic standpoint, enforcement grew lax, and program budgets for monitoring the status of natural resources were slashed.[18] The anti-environment strategy of the Reagan-Bush period was simple, direct, difficult to combat, and successful in slowing down or frustrating legislative intent. The technique was non-enforcement of the law, weak enforcement of the law, and perverse enforcement of the law. This is difficult stuff to combat effectively in court or in the political arena.

Despite these significant setbacks, however, the Reagan administration actually helped rouse public support for the environment, just as the Bush-Cheney administration did in pursuing anti-environment policies upon taking office in 2001. In the 1980s, Reagan's attempts to undo the work of the previous ten years galvanized the public to rally on behalf of the environment in greater force. In fact, membership rolls of the nation's major environmental organizations saw their biggest increases during this period.[19]

But business and industry—the primary targets of environmental reforms in the 1970s and 1980s—also grew more sophisticated. Businesses big and small alike became savvier at promoting their self-images to the public. At the same time, they worked to discredit environmentalists—a strategy designed to counter the negative publicity they encountered as environmentally unfriendly businesses. Many critics of the environmental movement soon found a home with the "property rights" and "Wise Use" groups.[20]

A chief weapon in the anti-environmentalists' arsenal has been to paint environmentalists as wild-eyed, radical extremists. Sometimes it has worked. Eric Freyfogle, a professor at the University of Illinois Law School, said the countermovement has succeeded, at least in part, in finding ways to distort issues of environmental protection into issues of "people versus the environment."

That occurred, unfortunately, in the well-publicized fight to save old-growth forests from logging in the Pacific Northwest. I argued at the time that we shouldn't allow the debate to become one of "jobs versus owls," but it was unavoidable. The only way to stop logging in the old-growth forests was to invoke the federal Endangered Species Act, which protected the spotted owl living in those forests. The whole forest ecosystem was threatened by logging, but only the spotted owl—classified in 1990 as "threatened" under the Endangered Species Act—was federally protected. Despite this broader environmental threat, the press went on to characterize the lawsuit as "jobs versus owls." A concept like this is easier to explain than ecosystem damage or the fact that jobs are lost when the resource base is destroyed. It also is more easily distorted, as the logging proponents showed.

Incidents like that, Freyfogle said, have helped make environmentalism "a dirty word" in this country. "Whoever thought of that, it was a wonderful rhetorical tour de force—and it was a disaster for the environment," he said. "When people say 'owls vs. jobs,' they're going to say jobs. If the conservation community were leading the rhetorical battle, it would have been 'short-term exploitation vs. long-term sustainable living on the land.'"[21]

Michael Kraft, a professor specializing in U.S. environmental policy at the University of Wisconsin–Green Bay, agreed that the countermovement has made some inroads against environmentalism but said it has been unable to reverse a more than thirty-year history of strong environmental support. "What they've done is create a situation where it's OK to be against environmentalists, as though you're distinguishing the really significant work from the flaming radicals. So it's easy to say in polite company: 'Oh, you're just a tree-hugger,'" said Kraft. "At the same time, we have polls saying the overwhelming majority of the public is sympathetic toward the environmental movement."[22]

Despite rhetoric to the contrary, Kraft noted that polls show only a small percentage of the public is sympathetic to the anti-environmental movement. "There are many more people who are now skeptical of the term 'environmentalist,' but I'm not convinced it means very much because it's an artificial effort to discredit environmentalism. It is part of what the opposition has to do because there are so few people who actually support their agenda," he said.

In the polls, surveys measuring the public's attitude toward environmentalists have yielded mixed results. A 1999 poll by the Wirthlin Report asked respondents whom they would they trust most to give them information about whether a local chemical spill posed a risk to public health. Only 4 percent said environmental groups, compared with 13 percent who said they would most trust representatives of the company responsible for the spill.

In a survey tracking attitudes from 1987 to 1994, however, a *Times Mirror* poll asked respondents how closely the label "environmentalist" fit them. In both years, more than 50 percent responded favorably to the label, with less than 20 percent responding unfavorably.[23]

"Greenwashing"

It is a common complaint among environmentalists and others that corporations are claiming to be "green" when they aren't. "That's greenwashing," these critics say.

Clearly, if greenwashing is occurring, the public forum is the place to address it, expose it, publicize it, attack it, or do whatever you please. Some people want to pass laws against it. Well, pretty soon you're interfering with free speech. Greenwashing is not all bad; in fact, it is a generally positive development. It shows that even the bad guys want to look green to the public. It is a true public relations success when even your worst opponents claim to share your environmental concerns.

Still, it's helpful to know that this strategy exists in the marketplace and to recognize it for what it's worth. At its root, greenwashing is an effort by business and industry to appear responsive to the public's environmental concerns, while generally continuing business as usual. The strategy manifests itself in a variety of ways:

Eco-marketing

Supermarket and store aisles today abound with examples of green marketing or eco-marketing, ranging from products such as paper towels and toilet paper to laundry detergents and household cleaners.

These products are billed as safe for the environment. The fact that companies now provide consumers with "greener" choices is a positive development, but the wary consumer may be unaware that neither the government nor trade associations regulate these labels. Thus, a label touting environmental benefits may be just a label. Furthermore, the creation of these products doesn't necessarily mean real concern for the environment on the seller's part. For some companies, selling green products is just another way to expand their market—even as they sell lines of the same product that don't profess to be environmentally friendly.[24]

On a bigger scale, we've all seen large advocacy advertisements in newspapers that put forth a corporate point of view. Attempting to influence public opinion by essentially writing their own "stories," corporations buy full-page ads in which they extol the environmental practices of their company or attack scientific findings that global warming is occurring or species are going extinct. By 1980, this form of corporate advertising totaled an estimated one billion dollars a year.[25]

Industry is also taking its message to the streets, borrowing an approach used by environmentalists for years. In a commentary in *Mining Voice*, a mining industry trade magazine, editor Jeanne Chircop ventures that "'activism' and 'activist' really aren't dirty words at all." Noting that Americans show a small but growing interest in environmentally friendly products, she exhorts miners to be vocal proponents of their work. "We need to get out there and be LOUD about all the great things the industry does for the environment, the economy, technology and, most of all, for people," Chircop writes.[26]

Classroom Influence

Aware that the shaping of public perceptions often starts early, corporations are taking their messages directly into the classroom via textbooks and other materials. These corporations prepare the materials and often distribute them free of charge to cash-strapped school districts.

Procter and Gamble, for instance, at one time produced and distributed "educational packages" to nearly seventy-five thousand U.S. schools. Distribution of this package ceased in the wake of complaints about bias and misinformation, however. For example, the materials defended the company's destructive clear-cut forestry practice as one that "most closely mimics nature's own processes."[27] In another instance, an activity book distributed by the American Coal Foundation trumpeted the fact that the increased carbon dioxide associated with global warming "makes plants grow larger." And in yet another example, a teaching kit produced by logging giant International Paper told students that cutting mature trees promotes "the growth of trees that require full sunlight."[28]

These materials influence our young people. A mining industry newsletter, boasting about its efforts to teach students about the benefits of mining, ran the following comments from a student at a North Carolina high school visited by mining representatives: "I learned that mining is cleaner than most people thought. Miners do good for the environment so the animals can roam freely in the non-polluted fields and people use these minerals every day," a student named Ricky wrote.

There is, in fact, a movement under way to discredit environmental education as it is taught to children in schools today. It is a movement led by vocal right-wing think tanks and corporate spokespeople who promote the notion that environmental education is taught with a one-sided, liberal bias. Heading up this movement is former Arizona political science professor Michael Sanera, co-author of a 1996 book that has become a bible of sorts for the new anti-environment movement: *Facts Not Fear: A Parent's Guide to Teaching Children about the Environment*. In it, Sanera asserts that environmental education materials are factually inaccurate and favor the catastrophic, worst-case scenario of environmental issues in an attempt to scare and depress children about the future.[29]

A former director of the Center for Environmental Education Research at the conservative Claremont Institute think tank, Sanera led an unsuccessful effort to eliminate federal funding for environmental education and has helped stall reauthorization of the National Environmental Education Act of 1990. His subsequent campaign attacking environmental education state by state has been more successful. The 1996 repeal of Arizona's environmental education mandate, for example, cut funding for environmental education and killed the curriculum in that state.

Like Sanera's book, the movement appears open-minded on the surface, offering to "plug the voids" in teachings about global warming, biodiversity losses, and the like by providing students with differing scientific views. The problem becomes apparent, however, when one learns that the sources of these scientific opinions are hardly unbiased—many of them have ties to industry groups and conservative groups.

Richard Wilke, a University of Wisconsin System distinguished professor of environmental education and past president of the North American Association for Environmental Education, said folks such as Sanera are trying to cloud children's understanding of environmental issues by promoting views that are outside the scientific consensus. "Mr. Sanera would like to see, as he would say, 'A more balanced picture' of the environmental issues presented," Wilke said. "But what he's really saying is that some of the views that are much in the minority relating to particular issues should get equal weight to those that the majority of scientists hold."[30]

Wilke, who has more than twenty-five years of experience in the field of environmental education, said surveys he's been involved with show students are optimistic, rather than fearful, about environmental problems. Other surveys find widespread public support for environmental education, such as a 2001 Roper Starch poll that found 95 percent of Americans favor it.[31] On the downside, Wilke said, research shows that significant gaps still remain in children's and adults' understanding of environmental problems, limiting their abilities to make informed decisions and leaving them vulnerable to distortions of fact.

National Audubon Society President John Flicker has likened today's critics of environmental education to a "modern-day Flat Earth Society": "They argue that basic scientific theories such as evolution, global warming, and the effects of pollution are so unproven that government should not require their teaching, or should require them to be taught in tandem with religious or political concepts like creationism."[32]

The idea that we should shield our children rather than teach them about the environmental challenges in their world is backward rather than forward-looking. Students, of course, should be made aware of disagreements within the scientific community regarding environmental issues, but not to the point that industry biases are used to muddy those issues where scientific consensus has been reached. Teaching our young people about environmental problems and potential solutions is a course that inspires constructive action rather than paralysis.

Spinning Science

In their search for "sound science," corporations often fund studies to pit against government and other publicly funded research, furthering public doubt about the veracity of the research coming from both sides and giving science a black eye in the process.

Corporations are entitled to conduct their own studies, and certainly some are legitimate. But companies also fund research that is meant to distort or confound the facts to benefit their product or industry. This is a growing problem for science and the environment, as corporations—hoping to avoid regulatory hurdles or expense— have the ability to pour enormous amounts of time and money into such studies. Companies can further tilt this uneven playing field by bringing out a team of consultants, experts, and researchers that far outnumbers the government or publicly funded researchers defending environmental laws and the public.

Although a study's source of funding doesn't guarantee the result, there can be an incentive for scientists to produce findings favorable to the company paying the bill. Additionally, researchers who produce results contrary to the company's position or welfare may see their studies kept under wraps, such as when tobacco companies muzzled early studies linking smoking to addiction and disease. Other companies exploit disagreement among scientific studies as a means of delaying action on environmental compliance. Their call for more and more studies in the name of "sound science" is often simply a call for delay until a law is changed, a new regulator takes over, or the public resources needed to pursue the matter run dry.

Corporate front groups play an important role by answering industry's call for sound science. These groups, now flourishing in the United States, are funded by major corporations and often have names that sound official—even pro-environment. Front groups provide a vehicle for industry-funded experts to pose as independent scientists. The official-sounding American Council on Science and Health, for example, receives money from corporations ranging from NutraSweet to Monsanto and defends the safety of saccharine, pesticides, and growth hormones for dairy cows.[33] The Global Climate Coalition comprises nearly fifty trade associations and corporations representing fossil fuel, automobile, and chemical interests, opposes restrictions on greenhouse gas emissions, and lobbies to convince Congress and the public that global warming is a myth. Likewise, the National Wetlands Coalition opposes implementation of U.S. Wetlands protections as "burdensome and ineffective regulation of public property." The group is funded by oil and gas companies.[34]

The Wake-up Call

The American public has made it clear it doesn't want progress on the environment undermined. Yet public concern for the environment has been seriously tested during the last thirty years. Recall the congressional battles of 1995. The conservative swing that began in the 1980s culminated in a Republican-dominated Congress attuned to industry's frustration with regulation; Republican leaders believed Americans had finally relaxed their vigil on environmental issues. To some degree, they had, thanks to a recession and a sense that the environment was taken care of following the 1992 election of President Bill Clinton and Vice President Al Gore.

It was a gross miscalculation, however. Republicans in the 104th Congress quickly sought to repeal and weaken some of the same laws that brought about major environmental gains. Environmental agencies and policies became prime targets of the conservative "Contract with America" and its call for smaller government and less regulation. In what has since been described as a stealth attack on the environment, anti-environment riders were attached to appropriations bills to avoid public scrutiny.

Environmental groups and the public rallied. In one instance, environmental groups collected more than a million signatures on an "environmental bill of rights," delivered to Congress in 1995. Astute politicians also responded, in the White House and across the aisle. Some Republicans broke rank, casting votes siding with the environment.[35]

Senate Democrats used the occasion of Earth Day 1996 to level a symbolic counterattack. Forty-one Senate members sent an Earth Day letter to Republican Senate Majority Leader Bob Dole and House Speaker Newt Gingrich, pledging to block attempted rollbacks of environmental, health, and safety protections.

Citing more than two decades of "remarkable bipartisan consensus on protecting the environment," the Democrats used the letter to publicize the actions of congressional Republicans. "We have seen bills that would broadly attack existing health and safety protections; bills to repeal the 1990 amendments of the Clean Air Act; bills that would roll back water quality standards and weaken enforcement of clean water laws; bills that would slow down and often stop cleanup of hazardous waste sites; and even bills to eliminate community right-to-know protections," the lawmakers wrote.

The mid-1990s saw a more vigorous assault on our environmental laws than I have seen in thirty years. It was, however, more a symptom of the larger, anti-government sentiment of the time than anti-environment fervor. I said then and believe still that there is no general endorsement or public support for destroying the goals and progress of the environmental laws established in the 1970s.

Renewed Vigor

American public opinion rallied in support of the nation's environmental laws. The incident showed, however, that public attention to environmental issues tends to be reactive rather than proactive. Absent an urgent call to action, public attentiveness to environmental matters is on unequal footing with the relentless pressure lawmakers face from determined and well-funded corporate interests pursuing a so-called economy-first agenda.

Nevertheless, a growing number of people aren't waiting for Washington and are taking the lead, working hard to bring positive changes to their communities and businesses. The sustainability movement in this country is growing, recognizing that a healthy environment is the foundation for a healthy economy and society.

Initiatives to foster sustainable development worldwide were a primary focus of the 1992 Earth Summit. U.S. leaders, too, are realizing the importance of sustainability, as demonstrated by the creation in 1993 of the President's Council on Sustainable Development. In a 1999 letter to President Clinton summarizing its findings, the council sounded a note of hope in the face of daunting environmental pressures in the United States and around the world. The council wrote:

America's challenge is to create a life-sustaining Earth, a future in which prosperity and opportunity increase while life flourishes and pressures on the oceans, Earth and atmosphere diminish. Even as we see evidence that damage to natural systems is accelerating, we also see individuals, companies, and communities working together to find solutions that work. We believe the United States can meet the challenge if all Americans strengthen their will and capacity to find agreement on important issues about our future.

We are convinced that America's economic prosperity can go hand-in-hand with a healthy environment and a high quality of life for its citizens. As the awareness of sustainable development continues to increase in Americans from all walks of life, we hope that a consensus will build in the land . . . that sustainable development is both right and smart for America.[36]

The phrase "sustainable development" is, understandably, confusing. If it is understood to refer to only that kind of development that is sustainable over the long term, however, the confusion disappears.

The Environmental Citizen

This is a good time for environmental groups and lawmakers to build upon the environmental concern percolating around the country today. But individuals have a role to play as well. A commitment to personal responsibility for the environment is needed now more than ever. There is no time for on-again, off-again interest in today's environmental challenges. The global temperature does not rise and fall with the stock market, nor do the fates of the plants and animals struggling to keep their place on the planet. Only by re-evaluating prosperity in real-life terms can we find the impetus to protect those things that make us truly rich in life.

Admittedly, the sustainability challenge is formidable. But the multiplicity of small, grassroots victories erupting here and there across the landscape are significant and encouraging. It is an indication that the grassroots movement—always hard to quantify, but unmistakable once rallied—is calling on our political leadership to take notice again.

We are at a major point of departure. We have spent the three decades since that first Earth Day in rehearsals. We've passed legislation. We've tackled what we thought were the easy and obvious problems. None of it was planned in any organized way.

After more than thirty years of discussion, debate, legislation, and education, there has evolved a new level of understanding and concern about what is happening in the natural world around us. The public is prepared and, in the end, will support those measures necessary to forge a sustainable society if the president and the Congress provide the kind of inspired leadership called for to meet the challenge.

We are now at dead center. The rehearsal time is over. The stage is set; the audience is willing to look at the play. But are the key actors prepared to perform? Have they read their lines? Do they know what it is they're supposed to do?

To crank up the political machinery for a move along the path to sustainability, someone must spark the engine. The president is in the best position to do this. He owns the bully pulpit; he is the nation's chief educator, the superstar. He is the only one who can command top billing in newspapers, on television, and on the radio, whenever he wishes.

Congress plays the other major role. It is time for the president and Congress to reach an agreement that sustainability is the challenge of our time and to design a plan of action for the future. It took three decades of effort to achieve the level of environmental stewardship we have today; it will take at least that long to move us within sight of the ultimate goal of sustainability—if we don't delay too long.

There is no room, nor time, for partisanship. The president and Congress should face this issue in a unified and cooperative way and should persist until we reach the goal. Most of us have many chances to do the right thing in our lives, but few among us are afforded the opportunity to be a key player in launching a program that will reverberate down through history as an act of vision and statesmanship. The president and Congress have the chance to make that rare decision. They have that golden opportunity to heed Bismarck's elegant observation: "The best a statesman can do is listen to the rustle of God's mantle through history and try to catch the hem of it for a few steps."

PART 4
An Environmental Agenda for the Twenty-First Century

7

Achieving Sustainability

The opportunity for a gradual but complete break with our destruc-
tive environmental history and a new beginning is at hand.

—1999

Our goal as a society is to successfully address the fundamental issue
of our time—the forging of a sustainable society. Every nation on the
planet faces the same challenge: the creation of a society whose activities do not exceed the carrying capacity of its resource base; that is
to say, a society that manages its natural resources in such a way that
their ability to support future generations is not diminished.

The massive grassroots demonstrations on Earth Day 1970 finally
forced the environmental issue onto our national political agenda.
Happily, in the years since, we have learned a lot and achieved a lot.
We have seen a reduction in our air and water pollution, eliminated
the use of DDT in this country, established a broad-based program
of environmental education, created a legal framework for protecting endangered species, and much more.

In the meantime, however, the leadership of both political parties
together with the president have, for years, pursued population policies
that will destroy the environmental achievements of recent years and
frustrate any efforts to forge an environmentally sustainable society—
by all odds the overarching challenge of the century. Current policies,
if continued, will double the U.S. population to more than 500 million
by around 2075, and to around 1 billion sometime in the next century.

As noted earlier, a doubling of the population will translate to
doubling the total infrastructure of the country—twice as many, airports, grade schools, high schools, houses, apartment buildings,

133

office buildings, and more. It will mean more traffic jams, more crowding, half as much open space per capita, and less freedom of movement, opportunity, and choice. Doubling our population will do all of this and much more.

At the United Nations' 1994 International Conference on Population and Development in Cairo, 179 countries endorsed a proposal that every country is responsible for stabilizing its own population.[1] The United States endorsed the proposal but has done nothing to keep its part of the bargain. In fact, indications are that the political leadership of the country, including Congress and the president, has no interest in reducing population growth.

Over time, we have learned that most environmental problems are either preventable or at least manageable. With this knowledge in hand, we now stand at the threshold of a golden opportunity to change the course of history.

This era marks the start of the environmental challenge of the future—the challenge of sustainability. The degradation we witnessed from the start of the twentieth century to the end was a warning sign. The world's oil supply may be nearing its peak at a time when demand is exploding to meet the increased auto, truck, ship, and train transport of a global economy. Most ocean fisheries have reached capacity production or are overstressed. Countless components of the plant and animal world are being dislocated and devastated. Water tables are falling on every continent as the population expands. Those knowledgeable of the key environmental facts know that we face a serious problem involving the total intricate ecological system that sustains all life.

Stewardship of the environment is a practical necessity, not a philosophical choice. Opinion polls reveal that the public is aware of these problems, is concerned, and wants solutions. But motivating people can be difficult. Many solutions require a change in behavior, and often denial is the easiest path. The lack of political leadership by our elected representatives is the biggest hurdle, however. Without vigorous and persistent leadership, the goal of sustainability cannot be achieved.

A way to make environmental problems appear less daunting is to relate them to our communities and convey their relevance to our daily lives—as they unquestionably are relevant. Practically, the citizens of this country can make a big difference for the planet by focusing here at home, where we have the most influence. As a nation of more than 280 million people living on land larger, richer, and more diverse in geography than all of Europe, we have a caretaker's responsibility to this land, to ourselves, and to the world.

Since the first Earth Day, much work has been done. In what has been a piecemeal approach to the environmental challenge, we have tackled some of the most popular and important environmental problems. Who among us would oppose the cleanup of our air and water?

It is too soon to pat ourselves on the back for a job well done, however. Even after thirty years of work and vigilance we have just begun to talk about sustainability—What is it? Why is it important? How do we achieve it?

Both the political establishment and the public must understand the impact of these questions and must support the steps necessary to move us on a path to sustainability. The leadership must come from the president and Congress, the only institutions with the legal authority and political clout to implement a program of sustainability. The third key player in this undertaking is the public, which must understand and support the goal of sustainability, or it won't happen. Over the four or five decades it will take to move close to this goal, Congress, the president, and the public will have to move together—at the same time providing leadership and setting an example for the rest of the world.

We can start almost anywhere. For example, we can begin the necessary transition from fossil fuels to solar energy. We can reduce air and water pollution to a level that is easily managed by nature. We can stop overdrafting the groundwater supply, depleting our fisheries, deforesting the land, poisoning the ground with pesticides, eroding the soil, degrading the public lands, urbanizing farmlands, and destroying wetlands.

We can do this and much more. One thing is certain—we cannot afford to delay fixing problems here at home while we wait for the rest of the world to act. As a nation, we have it in our power to do much of what is necessary to achieve sustainability. The longer we delay, however, the more we undermine the livable quality of the environment and the resource base that undergirds the economy.

Environmental Education

Some time ago during a trip to Georgia, I was asked to speak to about 700 third and fourth graders at a school about 60 miles outside Atlanta. After I had spoken for three or four minutes, the youngsters were excited to ask questions. One little girl was waving her arms, so I called on her. She told me of how a week earlier she'd come home from school to find groceries on the kitchen table. "I saw a can of tuna, and there was no dolphin on it," she said. The little girl went on to relate that she got her mother, and they drove back to the store to exchange the tuna for a can with a dolphin on the label—a sign the tuna was harvested in a manner that didn't trap dolphins in the nets.

That story and many like it demonstrate the evolution of an environmental ethic in our young people, an ethic developed by educating children about the interconnectedness of nature and our own influence on the natural world. When finally we have nurtured a generation of leaders imbued with a guiding environmental ethic, we will be able to forge and maintain a sustainable society.

The goal of environmental education in the school—as well as in the home—is to nurture a generation imbued with this ethic. Its purpose is to spark children's interest in the world and how it works. This isn't difficult. Children are born loving the out-of-doors and are fascinated by birds, bugs, and other animals and what they do. Tapping into that natural curiosity and building upon it to teach our youngsters about the natural world and how it works is one of the most important things we can do to help the environment in the future.

The reason is straightforward. A public unaware of environmental problems will not be concerned. A public unaware that its behavior puts life on our planet at risk won't change its behavior. A public that doesn't understand the basis of our environmental problems, or that can't effectively communicate them, won't be heard by decision makers for the many critical environmental decisions before us.

Environmental education is the tool. Knowledge about the natural world around us helps students understand how forests, wetlands, and even microbes in the ocean perform important roles to keep the planet running smoothly. It teaches them that when pieces of nature are disturbed or destroyed, there's a cost to be borne and that, in some cases, the price is too high.

Political Leadership

Environmental education takes place in the home, the school, the workplace, and the neighborhood. It can take place within the government when political leaders make it a priority to focus the nation's attention on environmental issues.

It is time now for our key political leadership to join in a nonpartisan effort and design a plan to improve public understanding about environmental issues. It took three decades of effort to get where we are, and it will take another four or five decades to get where we want to be.

The challenge is to forge a society that is economically and environmentally sustainable. Because sustainability is primarily a political challenge, we start with the two political institutions that share the key to the whole enterprise. Success or failure will turn on what kind of leadership comes from the president and Congress.

STATE OF THE ENVIRONMENT ADDRESS

Delivery of an annual State of the Environment address by the president to Congress, coupled with regular congressional hearings on sustainability, would initiate the kind of public dialogue that must precede all major decisions on controversial environmental matters.

There is a well-established tradition in this country of an annual presidential message to Congress on the State of the Union. There is, however, no tradition of a message on the state of the environment—despite the fact that the true "State of the Union" is almost entirely dependent upon the nation's underlying resource base. The economy ultimately will be destroyed by continued environmental degradation, yet stories about the Dow and the stock market are featured on the front page of newspapers and are the topic of conversation on the street every day. We have to get to the point where the environment is part of this daily conversation.

The State of the Environment message should be delivered every spring sometime after the traditional State of the Union address. In it, the president should outline environmental challenges that merit the nation's immediate attention and what challenges lay on the horizon. The address would be front-page news and, once established, would become a forum to provoke dialogue and discussion. In essence, such an address, along with congressional hearings, would serve as a national educational program.

Presenting Congress with this annual message on the state of the environment would set a powerful precedent that would be difficult, if not impossible, for future candidates and presidents to ignore. Likewise, vigorous leadership by the president on the issue of sustainability, in coordination with congressional hearings, would have historic consequences and would be an important legacy for any president to leave the nation.

CONGRESSIONAL HEARINGS

Congress is the other key player in this effort. Its primary and critical role must be a combination of educating and legislating. Again, public opinion polls show overwhelming concern for the environment and support for whatever measures may be required to maintain a clean environment. What particular measures may be required is not widely understood, however. Until it is, the public won't support—and Congress won't pass—the necessary legislation.

To enhance this understanding, Congress will need to conduct an extended series of congressional hearings, debates, and legislative actions involving the broad spectrum of issues that must be addressed along the way to sustainability. This may seem an onerous and intimidating challenge because it will extend over a considerable time and involve significant debate and controversy. The only rational choice we have, however, is to begin the process without delay.

To make this undertaking succeed will require a sustained, cooperative, nonpartisan effort unlike any other in our peacetime history. The state of the environment, and how it affects the economy and our quality of life, must be better understood by the public and our political leaders. Congressional hearings—held once or twice a month over the next several years, preferably by a joint congressional committee—could accomplish this.

Conducting these hearings would require about four to eight hours a month, certainly not a burdensome task. It's been done before. For example, from 1969 to 1979 I chaired the Senate Small

Business Committee when it conducted 135 hearings on prescription drugs—their pricing, uses, and abuses. Within a year the hearings were receiving national coverage from the three television networks and the wire services. The message got out.

Of necessity, these sustainability hearings must cover a range of significant issues spanning the environmental spectrum. That means exploring topics such as how we:

- make the transition from our heavy reliance on fossil fuels to a significant reliance on solar energy;

- move to restore ocean fisheries;

- reduce air and water pollution to a level manageable by nature;

- preserve our magnificent heritage of public lands;

- shrink our excessive reliance on pesticides;

- stop overdrafting groundwater and reduce soil erosion; and

- preserve wetlands, forests, and biodiversity.

These and dozens of other issues will command attention.

Contrary to what one might hear from political naysayers, it is not a difficult task. Invite the witnesses, and a national discourse is under way. If this discourse is not initiated, it will delay by years and years addressing the question in any significant way.

The First Hearing: Preserving Public Lands

Where do we begin? Congressional hearings on sustainability could start almost anywhere. My choice, however, would be public lands, because almost everyone has some familiarity with national parks, national forests, wildlife refuges, or federal Bureau of Land Management lands.

These lands are a rare and remarkable heritage of almost 1 million square miles totaling about 26 percent of the U.S. land mass. No other nation on Earth has set aside such a vast mosaic of mountains, wetlands, lakes, rivers, seashores, islands, plains, forests, grasslands, and deserts. Within these bounds a sample of almost every major American landform is represented, and these lands are the only large expanses of natural areas left in the lower forty-eight states. Here are lands that would be recognized by our forefathers, lands inhabited by wildlife that cannot survive elsewhere, lands offering a rare condition of quiet undisturbed by the sounds of our technological progress, and immense vistas of scenic beauty not found elsewhere.

Early in the twentieth century, when the national park system was new in this country and unknown in any other, James Bryce, an Englishman, characterized it as "The best idea America ever had." Yellowstone National Park, the world's first national park, was created in 1872. Since then, more than a hundred countries have established national parks.

Our national park system was formally established by the 1916 Organic Act and now encompasses some 80 million acres. The act specified that the national parks be managed with the purpose of conserving "the scenery and the natural and historic objects and the wildlife therein." It further called for the parks to be managed in a manner that "will leave them unimpaired for the enjoyment of future generations."

How wonderful it would be if the park system were managed in compliance with the spirit and letter of the law. Sadly, it is not. Over many years a succession of presidents and Congresses have defaulted on this responsibility and have permitted all kinds of incompatible activities to proliferate, to the detriment of our parks. Obviously, the same activities that adversely affect wildlife, pollute the air, destroy the peace and quiet of the parks, and otherwise degrade the enjoyment of these special places violate the mandate we have been charged with carrying out.

Our entire national park system is in varying degrees of decline. As mentioned earlier, the sheer number of annual visitors to these parks is placing them in jeopardy. Snowmobiles are causing air and noise pollution in Yellowstone. At Yosemite, several thousand visitors stay in cabins and tents, creating a virtual city that has been described as "looking like downtown Los Angeles at midnight." In Grand Canyon National Park, 100,000 commercial tourism flights a year fly down the canyon, disturbing wildlife and the peace and quiet of that long-revered place. In 1985, Governor Bruce Babbitt of Arizona testified that the noise in the canyon was "equivalent to being in downtown Phoenix at rush hour." Contrast that with what Zane Gray wrote about the Grand Canyon in 1906: "One feature of this ever-changing spectacle never changes: its eternal silence."

This is just a snapshot of what is happening to the crown jewels of our public lands. At the current rate of degradation, the national parks as we know them will be gone within the next three decades. If we continue with business as usual, one can make a strong case that thirty years from now the national parks and forests will be reduced to noisy theme parks or amusement parks. Our national forests and Bureau of Land Management lands are under a similar siege but are even more threatened because they lack the legal protections created to guard the national parks. These lands are being degraded by four-wheel-drive vehicles, motorcycles, snowmobiles, and jet skis.

This abuse is targeted at more than a quarter of our nation's land base, a fact that, in and of itself, should justify extensive hearings to inform Congress and the public what is happening and what is at stake.

There are dramatic things we need to do, and it will be up to the public to support any measure that curbs or eliminates activities that harm our public lands. For starters, the use of off-road vehicles on public lands should be phased out; cattle grazing should be re-evaluated and either reduced or phased out wherever it compromises the resource base.

This is the stuff of controversy, however, and it begs a national public discussion.

The Second Hearing: Population Growth

If we accept that forging and maintaining a sustainable society is the critical challenge for this and future generations, we must also accept that stabilizing our population will be key to determining our success or failure.

The most pressing issue of all for this country is the matter of what America will be like when our population of more than 280 million doubles to more than a half-billion in the next seventy to seventy-five years—a rate at which we will join China and India in the billion population class sometime in the next century. Clearly, population is a world problem, and many nations will suffer from overpopulation problems worse than those facing the United States. But we have a problem here at home, and if we wait for the rest of the world to come along and do what's right, we'll join the less-industrialized nations as a nation of degraded, overpopulated lands.

The need for the countries of the world to control population growth before it causes irreversible damage to the environment is backed by a number of respected, independent authorities—plus common sense. Recall that population control was central to discussions at the UN's 1994 International Conference on Population and Development. Two years earlier, in his 1992 book, *The Diversity of Life*, Nobel laureate E. O. Wilson voiced a similar opinion. "The time has come to speak more openly of a population policy," he wrote. "By this I mean not just capping the growth when the population hits the wall, as in India and China, but a policy based on a rational solution of this problem." Wilson added, "The matter should be aired not only in think tanks, but in public debate."

More recently, a task force of the President's Council on Sustainable Development made a similar case for population stabilization in its study of U.S. population and consumption trends. In a 1996 report, the task force warned that continued population growth adds to the nation's burdens to reduce poverty, improve education, and provide health care for all Americans. It also warned that continued growth could thwart efforts to provide new and higher-quality jobs for Americans and to improve wages for U.S. workers. "In short," the task force wrote, "the United States is already severely challenged by the need to provide better opportunities for millions of disadvantaged citizens, and continued population growth will exacerbate those challenges."[2]

The panel also took on a popular argument that a steady influx of human capital is needed to sustain the nation's growing economy, citing the 1972 findings of a U.S. commission headed by John D. Rockefeller III that was similarly assigned to assess the costs of the nation's population growth. That study, by the President's Commission on Population Growth and the American Future, examined the relationship between population and prosperity—comparing the effect of an American population averaging two children per family to one averaging three children per family. "The nation has nothing to fear from a gradual approach to population stabilization. . . . From an economic point of view, a reduction in the rate of population growth would bring important benefits," the commission wrote. It concluded: "We have looked for, and have not found, any convincing economic argument for continued population growth. The health of our country does not depend on it, nor does the vitality of business nor the welfare of the average person."

More than 80 million people and thirty years after the five-volume Rockefeller report was released, we have yet to begin a serious discussion of the critical issues it raised. The reason is simple enough: the commission was treading on sensitive territory.

Since the first Earth Day, most Americans have accepted the notion that overpopulation exists, and they show little desire for an ever-larger U.S. population.

Back in 1971, a poll by the U.S. Commission on Population Growth and the American Future found 22 percent of those surveyed said the country's population should be smaller than the nearly 200 million residing in the United States at that time. In 1974, nearly 90 percent of respondents to a Roper poll said they weren't interested in seeing more people living in the United States. By 1995, a Roper Starch poll reported that more than half of American adults surveyed said the nation's population was too large, and a Gallup poll that same year reported nearly two-thirds of Americans said the United States was at the point of overpopulation.

When I was born in 1916, the U.S. population was about 98 million. By the time the United States entered World War II in December 1941, the population was about 132 million. When the first Earth Day took place in 1970, the count was 200 million.

Tremendous ecological damage occurred as a result of this growth. In the United States, wetlands were drained, farm soils were swept into streams, and pristine rivers were dammed. Worldwide, we filled the skies and waters with chemicals, rechanneled rivers, and extinguished many species. And we didn't stop there. In the second half of the century, we added even more powerful technologies and used them recklessly across even larger parts of the planet's surface. The environmental destruction far surpassed the mind-boggling destruction of the previous half-century.

In 2000 the U.S. population topped 280 million. Not surprisingly, adding population hasn't improved American society, the economy, or the environment. Yet we are headed, at current growth rates, toward having well over 500 million people on the same land resource within the next seventy-five years and 1 billion people within the next century. Does anyone imagine we can grow like that without tremendous cost to the environment and our quality of life?

The U.S. birthrate is at replacement level, or about 2.1 children per woman on average. This birthrate would bring about population stabilization over a relatively short time. Yet we won't stabilize our population as long as immigrants to the United States continue to add 1.3 million people to the population each year—300,000 of them entering the country illegally. It is a fact that until we address this growing influx of immigrants, who account for about one third of our annual population growth, the population will continue to grow indefinitely despite the nation's success at achieving a replacement-level birthrate.

Never has an issue with such major consequences for this country been so ignored. Never before has there been such a significant failure by the president, Congress, and the political infrastructure to address such an important problem. We are faced with the most important challenge of our time—the challenge of sustainability—and we refuse to confront it. It is the biggest default in our history.

The reason for this silence is simple. In order to bring a halt to exponential growth, the number of legal immigrants entering this country would have to match the number of emigrants leaving it— about 220,000 people per year. Yet, while federal actions have increased the immigration rate dramatically during the last four decades, any suggestion that the rate be decreased to some previously acceptable level is met with charges of "nativism," "racism," and the like. Unfortunately, such opposition has silenced much-needed discussion of the issue—recalling the political smear tactics of the late Senator Joe McCarthy. The first time around it was "soft on communism." This time the charge is "racism" because a significant number of immigrants are of Hispanic descent. Demagogic rhetoric of this sort has succeeded in silencing the environmental and academic communities and has tainted any discussion of population-immigration issues as "politically incorrect." As frustrating as it is to see the president and members of Congress running for cover on such a monumental issue, it is nothing short of astonishing to see the great American free press, with its raft of syndicated columnists, frightened into silence by political correctness.

The issue is not racism, nativism, or any other "ism," however. The real issue: numbers of people and the implications for freedom of choice and sustainability as our numbers continue to grow. Population stabilization will be a major determinant of our future, how we live and in what condition; talk of it should not be muzzled by McCarthyism or any other demagogic contrivance. Rather, the issue must be brought forth and explored in public hearings and discussions precisely because it is a subject of great consequence.

Noting that the United States consumes more resources than any other country in the world, the task force of the President's Council on Sustainable Development warned of "enormous implications" for the environment, economic progress, and quality of life if U.S. population growth continues at current rates or higher. "Coupled with the technologies and resource consumption patterns that underlie the U.S. standard of living, population growth in America produces an environmental impact unparalleled by any other country at this time," the task force wrote in 1996.

Adding more and more people to the U.S. population base will do nothing to relieve the overload in this country, nor will it significantly relieve the overload on those nations sending immigrants. Indeed, if we don't set limits and stabilize our population, we'll continue to grow until we reach the point where conditions are as bad in the receiving countries as in the sending countries. For their part, overpopulated nations will fail to take the steps necessary to protect their own natural resources, and the global environment will be increasingly stressed.

As Garrett Hardin, an author and professor emeritus of human ecology at the University of California, has said: "Admitting immigrants from overpopulated countries amounts to taking on their problems which they haven't solved. If we take on their problems, they will never solve them by other means."

The fact that lower birth rates are found in industrialized countries where women have access to education, job opportunities, and family planning services means we should do everything we can to provide education, family planning, economic advice, and technical aid and assistance to those countries where such opportunities are lacking.

But we also must consider the causes of the refugees' flight and see therein the environmental and social conflicts that often accompany overpopulation. Many immigrants to the United States are refugees because environmental problems are not being dealt with in their native countries. For example, a large number of immigrants in recent years are from places with diminishing croplands—Central and South America and the Philippines, for example. Others come from nations such as India, where the ecological systems have been ravaged.

Indeed, many of the world's violent conflicts are heavily influenced by—if not caused by—overpopulation and environmental mismanagement of agriculture, water, and forestry resources. Environment and overpopulation problems often cause these resources to be in short supply, a tenuous situation that leaves the populace vulnerable to those using ethnic and religious differences to gain power.

We don't often think of warfare, civil war, and violent internal conflicts as being caused by excessive population. The political interpretation and media reports in the United States typically characterize these conflicts as religious, ethnic, tribal, or peasant disputes, or the result of some failing in the moral status of the cultures in which they occur. Were there not overwhelming environmental problems in these nations, however, many of these conflicts would be smaller or wouldn't have happened at all.

That population will stabilize sometime is inevitable. Unfortunately, if left to its own devices, this stabilization will occur when crowding, crime, inconvenience, noise, polluted air, congestion, and food and water shortages confront us wherever we turn.

Environmental Education: In the Classroom and Beyond

The effort to achieve sustainability and improve public understanding of environmental issues must take place at more than the federal level, however. It must occur in the states, in local school districts, and in homes—where the greatest number of citizens can participate and shape environmental education.

Positive changes already are under way in many schools. In the years since the first Earth Day, I've spoken to a number of grade school and high school audiences every year, and I have seen progress. The young folks—grade school and high school students—are interested in and excited about the environment and environmental causes. And the teachers tell me their students love to be involved in environmental issues, whether it's planting trees, creating little wetlands, or planting native grasses and flowers.

Today's seventh and eighth graders also ask more sophisticated and penetrating environmental questions than college seniors asked back in 1970—clear testimony to the success of environmental education in the schools. Thousands of schools have environmental science classes these days, and many more blend some type of environmental education into subject areas ranging from English to social studies. My home state of Wisconsin in 1984 became the first state in the nation to mandate environmental education for grades K-12, and dozens of states now feature environmental education initiatives ranging from new teacher-training requirements to statewide assessments of environmental literacy.

At home, many parents already teach basic environmental behaviors to their children, helped by television and the Internet, which offer captivating nature documentaries and Web sites.

Convinced that environmental education would always be critical to the country's future environmental health, I authored the first National Environmental Education Act with the hope that the federal government would give states and local school districts the tools to instruct young people about the planet and how it works. Signed into law by President Richard Nixon in 1970, the act established the Office of Environmental Education, placing it in the Department of Health, Education and Welfare. The new office was charged with awarding grants to develop environmental education curricula and provide professional development for teachers. But the department wasn't interested in environmental education, nor was the Reagan administration, so the program lapsed for lack of funding.

Fortunately, Senator Quentin Burdick (D-North Dakota) led the 101st Congress to pass the National Environmental Education Act of 1990, re-establishing the office, but this time placing it within the EPA. In reviewing the legislation, the House Committee on Education and Labor stressed the importance of increasing public understanding of environmental issues and said environmental education is an investment in preventing costly environmental disasters and promoting the sustainable use of natural resources.

More must be done, however.

The National Environmental Education Advisory Council warned in a November 2000 report to Congress, "The overall environmental education effort is currently too diffuse and fragmented to effectively fulfill its critical role in our society." Specifically, the council reported that although progress has been made since 1990, the overall national effort to promote environmental education is far weaker than it should be in terms of adequate funding, evaluation tools, and coordination and use of resources. "It appears that neither the general public nor the corporate world fully realizes or appreciates the link between environmental education and a sustainable economy," wrote the eleven-member panel of citizens and experts.[3]

The council issued another powerful statement in which it underscored the critical role of the individual in maintaining the nation's environmental health. Noting that the era of relatively easy environmental cleanups targeting industry through regulation and enforcement is now behind us, the council highlighted environmental education as a tool to bring about voluntary behavioral changes from the public.

"Research shows that the largest source of pollution today is nonpoint source pollution caused by the collective behavior of individual citizens. Reliance on our cars, insistence on perfectly shaped fruits and vegetables, and rejection of environmentally responsible behavior that inconveniences us in any way are all contributing to a serious degradation of human and environmental health," the council wrote. "Regulation represents the lowest level of the public's commitment to a quality environment. . . . We need a more sustainable and higher level of commitment from our citizens to secure our collective environmental, economic and physical health."

To accomplish this, the group offered a number of solid recommendations. First and foremost is reauthorization of the 1990 National Environmental Education Act with adequate and consistent annual funding. The council's other recommendations included that we:

- Create sustainable funding sources for environmental education, such as establishment of a national trust fund and federal block grant program.

- Require environmental education concepts and skills in state academic standards and assessments and include environmental education concepts and skills on exit exams for teacher certification.

- Strengthen the evaluation and assessment of environmental education.

Nurturing a public that is willing to inform itself about environmental issues, engage in debate, and weigh alternative solutions is at the heart of what environmental education can and must accomplish. Until the public, the politicians, and the corporate world understand the link between economic growth and a sustainable environment, such informed debate will not occur.

Environmental Education in the Community

From a practical viewpoint, individual action is essential to raise our national consciousness about the Earth and our impact on it.

Education is not primarily a national policy effort, despite the discussions we have about it in presidential election years. Rather, most education occurs by local initiative and under local control. So for those who ask "What can I do to advance environmental education in the United States?" some specifics follow:

- Work at the local level to ensure environmental education is part of the curriculum of schools in your community and state. Guidance on this can be obtained from the North American Association for Environmental Education. Work through your parent-teacher organization and in cooperation with local environmental educators. Contact your local school board members to encourage their attention to environmental education.

- Contact your state public instruction department and urge their support for expanded environmental education. Many states have environmental education organizations.

- Support your local nature center, many of which are private and need your financial contributions and volunteer services.

- Work with your state environmental department or state park education program. Take your family to programs that teach about your local environment and its challenges.

The Environmental Ethic

When we think about urgent environmental issues, we think of such issues as global warming, pollution of the oceans, acid rain, hazardous wastes, and exponential population growth. There is, however, an overwhelming case to be made that lack of a pervasive environmental ethic in our culture is our most serious conservation problem.

For two hundred years we have been guided by a growth ethic that has promoted unparalleled exploitation of all our natural resources without regard for environmental consequences. Future generations now face a huge environmental deficit they may never be able to pay off.

Yet some argue it would cost too much and jeopardize economic growth if we were to adopt measures to prevent further environmental degradation and begin the difficult process of restoration. What these critics fail to recognize, however, is that our long-term economic health has been dangerously compromised by our policy of degrading and consuming our natural resource base and charging it to future generations. We are dealing with a politically bankrupt, slash-and-burn way of thinking that is economically, environmentally, and morally indefensible. The challenge before us now is to alter our conduct in significant ways, both as a nation and as individuals.

If we are to attempt reversing two centuries of irresponsible resource exploitation, we must nurture a generation imbued with a guiding environmental ethic that prompts each individual to always ask: If we intrude here, what will be the environmental consequence? If we had always asked that question in the past, much of the environmental damage we have caused would not have happened.

Consider that state highway departments all across the nation have engaged in the massive destruction of valuable wetlands to punch roads through the middle of marshes in order to save a few seconds of travel time. In almost every case, the 15 or 20 seconds saved aren't worth the environmental cost.

Every week we witness similar shortsighted actions by local planning boards and individual landowners. Every day we observe the corruption of environmental and public health science by special interests. Every hour we see short-term, anti-environmental actions by corporations, as well as by state and federal governments. Waivers are granted to exempt shopping malls, housing subdivisions, and a host of other land uses from reasonable environmental planning. Air and water emissions from factories are not reported accurately, and many chemicals are not adequately tested before they are introduced into the food chain. Federal facilities are allowed to ignore pollution cleanup procedures. Our own tax dollars are spent to subsidize private companies involved in environmentally destructive mining and timber activities on public lands, while in many states environmental regulators become the servants of business interests that contributed to the governor's last campaign. Absent a guiding environmental ethic, our culture will continue to destroy enduring natural values for a handful of silver today.

Aldo Leopold's Land Ethic

Most discussions on environmental ethics begin with reference to naturalist Aldo Leopold's land ethic. As he put it, "Quit thinking about decent land use as solely an economic problem. Examine each question in terms of what is ethically and esthetically right, as well as what is economically expedient. A thing is right when it tends to preserve the integrity, stability, and beauty of the biotic community. It is wrong when it tends otherwise."[4]

Some economists are quick to say that economics should govern all decisions. They reduce life to a series of decisions theoretically serving the individual. The fact is this: you get what you measure. If you measure the nation's well-being by counting the output of widgets, the nation will focus on putting out more widgets. The blindness of economists lies in measuring the output of goods and services with no attention to the damage done by that output. Most economists do not examine how long and well our planet's oceans, forests, and food chain can biologically sustain this process. Like the shortsighted shop owner, they worry only about the balance in the till day by day.

Economists also give little or no thought to the price we pay when something is lost. The fallacy is that substitutes are easy to find—if we use up the oil, we'll replace it with nuclear energy or coal; if we destroy the rainforests, we'll get along without them; if the frogs disappear, nature will find a way for the other creatures to carry on. In the physical world, however, substitutes are not true substitutes. The reality is that each of these losses comes at a cost to the environment, the economy, and our society.

The Environmental Ethic and Religion

Religious people from all traditions have long agreed with the concept of environmental responsibility. Protestant, Catholic, Orthodox Jew, Muslim, Buddhist, and Hindu—all believe Earth is part of God's creation.

Paul Gorman, executive director of the National Religious Partnership for the Environment (NRPE), describes the faith community's renewed interest in environmental protection as a reawakening to religious teachings and traditions that long preceded Earth Day and the modern environmental movement. Moreover, Gorman noted that care for the Earth is inextricably linked with religion's humanitarian focus: "When you're focused on human well-being, it requires understanding that economic well-being means environmental well-being," he said.[5]

An alliance of major faith groups and denominations in the United States, the NRPE was formed in response to a powerful international appeal in which thirty-two Nobel laureates and other eminent scientists called on the global religious community to make a commitment "to preserve the environment of the Earth."

"Today, suddenly, almost without anyone noticing, our numbers have become immense and our technology has achieved vast, even awesome, powers," the scientists wrote in their 1990 appeal. "Intentionally, or inadvertently, we are now able to make devastating changes in the global environment—an environment to which we and all the other beings with which we share the Earth are meticulously and exquisitely adapted. We are now threatened by self-inflicted, swiftly moving environmental alterations about whose long-term biological and ecological consequences we are still painfully ignorant. . . . We are close to committing—many would argue we are already committing—what in religious language is sometimes called Crimes against Creation."[6]

Noting that the answers to some of today's most pressing environmental crises—such as a voluntary curbing of population growth and conversion from fossil fuels to nonpolluting energy sources—will encounter widespread public denial, resistance, and inertia, the scientists appealed to the ability of religious leaders to guide personal conduct.

This unprecedented appeal did not go unnoticed. A response signed by 271 spiritual leaders from 83 countries, including 37 heads of national and international religious bodies, welcomed the appeal as "a unique moment and opportunity in the relationship of science and religion."[7]

In the years since, religious leaders and environmentalists have engaged in several coordinated efforts to educate the public about such complex environmental issues as global climate change. The joining of the religious community with the scientific community on the matter of planetary health is encouraging, for it will take people of all faiths and persuasions to bring us along the path to sustainability.

Gorman said the faith community shares common goals with environmentalists and scientists when it comes to the environment, but the group approaches stewardship of the Earth from a different perspective. "We're about transformation, not legislation. We're about values, not regulations. We're about principles, not particulates," he said. It is a perspective and voice sorely needed in our move toward sustainability.

The Environmental Ethic and Technology

Sometimes our blind faith in technology gets us into trouble. So it was with pesticides in the 1960s, and so it may be with genetically modified foods and biotechnology today.

The multinational chemical, pesticide, and seed companies tell us they're going to feed the world's growing population with bigger and better crops that have been boosted by biotechnology. But we should exercise caution before agreeing that they have a panacea. There are too many questions and too few answers. Proponents say they can modify natural living organisms without problematic effects. Yet there are a number of red flags scientists are raising as this technology moves along:

- The cumulative and synergistic effects of chemicals and pesticides reacting with one another.

- The potential for genetically modified crops to contaminate organic and other nonmodified crops.

- The unintended effects on other species.

Clearly, more study is needed to determine whether genetically modified foods are safe. We are toying with nature, and we haven't sufficiently evaluated the implications. Several decades ago many people raised similar concerns about pesticides, warning that we were not yet aware of the long-term ramifications of their use. We are still learning today.

Scientists now are pushing technology further, blending pesticides into corn and fish genes into tomatoes—jumping species barriers and growth limitations that have been in place for thousands of years, probably for a reason.

Other researchers talk of extending human life expectancy to anywhere from 150 to 200 years and already have made it possible for a woman to bear a child long past her normal reproductive years. These developments and others aimed at extending and expanding human life on Earth are moving forward with apparently little thought of the consequences to our already overtaxed planet.

Science is a wondrous tool that has helped the human species achieve comforts and goals beyond our ancestors' wildest imaginings.

It has helped us beat the odds on many fronts, enabling us to break the sound barrier, defy gravity, and even cheat death. Clearly scientists don't perform their work in a vacuum. Their results profoundly influence life on Earth as we know it—as the scientists who invented the atomic bomb in the last century knew, and as scientists working to unlock the mysteries of DNA and stem cells know now.

Some scientists will say their research is conducted in the name of science and the pursuit of knowledge, and that this is justification enough. But the scientific community has a greater responsibility to the world we all share. The public also deserves to be, and must be, a part of this discussion, for it will either reap the rewards or bear the brunt of changes made in the name of technological progress.

The biotech genie is coming out of the bottle. Before it's all the way out, let us pause to be sure we understand the ramifications of tinkering with nature in this way.

Gathering Momentum

We are not starting from zero. The majority of Americans support the concept of environmental stewardship. Public opinion polls conducted in dozens of nations, rich and poor, Western and non-Western, show that environmentalism has made headway. We have indeed come a long way since that first Earth Day in 1970. We have the first building block.

A second building block is being moved into place, and it's interesting to note how it has come about so quickly on a global scale. Issues of sustainability and population took hold at the United Nations Conference on Environment and Development in Rio de Janeiro in 1992. Since then, the sustainability movement has gathered momentum around the world. In less than a decade, grassroots work to make everything from neighborhoods to industries to cities more sustainable has blossomed. Some of this labor is just beginning to bear fruit. Other efforts will take more time.

But a series of special success stories of the past decade is exemplified in the power of the people. They're taking matters into their own hands, from Chattanooga in Tennessee to Curitiba in Brazil. They're resurrecting waterfronts, recycling buildings, and reclaiming industrial wastelands. They're cleaning rivers, planting trees, and restoring wetlands. These folks aren't waiting for Washington or the World Trade Organization or even their statehouses to figure out what must be done. They're at work, in small groups meeting in churches and large networks communicating over the World Wide Web, figuring it out and making it happen.

The neighbor, the parishioner, the student—all are living examples of the Earth Day motto adopted more than a decade ago, "Think Globally, Act Locally." That motto, catchy and apt at the time, may need some refining as we try to focus in on what must be done now to encourage us in our daily actions to heed and live by an environmental ethic.

When we think of the world our grandchildren will live in fifty years from now, we can only hope that the work of those grassroots groups multiplies exponentially across the nation and the world. If we are really to start behaving the way we preach, we must keep our children and grandchildren at the forefront of our minds when we consider how we live, what we leave, and whether we care enough to take action or leave that to others.

The Challenge of Sustainability

We are dealing with a social, ecological, and economic challenge unlike any other in our history. It is a challenge that begs for the kind of dedicated, inspirational leadership provided by Franklin Roosevelt and Winston Churchill in their pursuit of victory in the Second World War. Yet this challenge is far more serious than the military threat to the democratic West in World War II. Nations can recover from lost wars—witness Germany and Japan—but there is no recovery from a destroyed ecosystem.

We are pursuing a self-destructive course. We are fueling our economies by consuming our capital—that is to say, we are degrading and depleting our resource base and counting it on the income side of the ledger. Clearly, this is not sustainable over the long term. The hard fact is that while the population is booming here and around the world, the resource base that sustains the economy—and all of the Earth's creatures—is dwindling. It is not just a problem in faraway lands; it is an urgent, indeed critical, problem here at home right now.

Intellectually, we finally have come to understand that the wealth of the nation is its air, water, soil, forests, minerals, rivers, lakes, oceans, scenic beauty, wildlife habitats, and biodiversity. Take this resource base away, and all that is left is a wasteland. In short, that's all there is. That's the whole economy. That's where all the economic activity and all the jobs come from. These biological systems contain the sustaining wealth of the world.

These same systems are under varying degrees of stress and degradation in most places around the planet, including the United States. As we continue to degrade them, we are consuming our capital. In the process, we erode living standards and compromise the quality of our own habitat.

It is a dangerous and slippery slope.

We are not just toying with nature. We are compromising the capacity of natural systems to do what they need to do to preserve a livable world.

We have evolved willy-nilly into a frenzied, consumer-oriented, throwaway society, and in the process we are dissipating our sustaining resource base.

Reaching a general understanding that sustainability is the ultimate issue will finally bring us face-to-face with the political challenge of forging a sustainable society during the next few decades. It is a challenge we can meet if we have the leadership and the political will to do so.

Three decades ago the public saw a similar challenge and launched a battle to reclaim the natural world. Today, a new restlessness brews in the land. Many of our citizens are seizing upon the goal of sustainability and challenging our leaders to do the same.

All of this will be enormously complicated and controversial, far beyond anything ever before attempted. Yet the debate and controversy are vital to the process of developing public understanding and support for making the hard decisions. If we fail to make the right decisions, nature will make them for us and for all future generations.

All of us must do our part, and we must demand that our leaders do their part as well. If we can accomplish this, there is no reason to fail.

An Appeal

Earth Day owes its success to the millions of people around the nation who rose to the occasion on behalf of the planet. But we owe a special debt of gratitude to that 1970s generation of young people—grade school, high school, and college students—who supplied the energy, enthusiasm, and idealism that contributed so much to that first celebration. And now, what a wonderful opportunity the next generation has to build on that legacy. Indeed, if this generation of students can persuade the political establishment to initiate a national dialogue on sustainability, it will mark another turning point in our history.

The youth of America must be vigilant in guarding against those who would erode or erase the hard-won environmental progress in this country. Special-interest groups will always attempt to weaken the laws protecting the air and water, the wildlife and land, when in fact these laws should be strengthened. Already, inroads have been made that threaten to undermine some of the most fundamental environmental gains of the last thirty years. Such assaults on the nation's environmental achievements are continuous; our defense must be unwavering.

This is a special appeal to the youth of America, without whom Earth Day would not have achieved what it has achieved, and without whom the new challenge of creating a sustainable world cannot be met. It is an appeal to America's youth to pick up where their parents left off. It was young people, after all, who embraced Earth Day with the greatest enthusiasm in the months leading up to that first celebration. And it is young people who help keep the tradition alive today, learning about and celebrating their planet in kindergarten classrooms and on college campuses across the nation.

Today's generation of young people and the generations that follow will reap the rewards or suffer the consequences of the critical decisions we make during the next several decades. It is their future, and their children's future, at stake.

Appendix 1

Letter to John F. Kennedy

Regarding the president's proposed resources and conservation tour

LISTER HILL, ALA., CHAIRMAN

WAYNE MORSE, OREG.
RALPH YARBOROUGH, TEX.
JOSEPH S. CLARK, PA.
JENNINGS RANDOLPH, W. VA.
HARRISON A. WILLIAMS, JR., N.J.
CLAIBORNE PELL, R.I.
EDWARD M. KENNEDY, MASS.
GAYLORD NELSON, WIS.

JACOB K. JAVITS, N.Y.
WINSTON L. PROUTY, VT.
PETER H. DOMINICK, COLO.
GEORGE MURPHY, CALIF.
PAUL J. FANNIN, ARIZ.
ROBERT P. GRIFFIN, MICH.

STEWART E. MC CLURE, CHIEF CLERK
JOHN S. FORSYTHE, GENERAL COUNSEL

United States Senate

COMMITTEE ON
LABOR AND PUBLIC WELFARE
WASHINGTON, D.C. 20510

August 29, 1963

The Honorable John F. Kennedy
President of the United States
The White House

Dear Mr. President:

Some time ago you suggested I might send you some ideas respecting your proposed resources and conservation tour. Along with this letter I have enclosed several pages of quotations, some of which may be fitting for your speeches.

Though it is likely most everything I might suggest has been considered, I will toss in all that occurs to me in the hope some of it might be useful.

163

The fact that you are going on a nation-wide tour will command great attention for several reasons including the fact that no President has done exactly this before. The question is how to maximize the effect -- how to hit the issue hard enough to leave a permanent impression after the headlines have faded away -- how to shake people, organizations and legislators hard enough to gain strong support for a comprehensive national, state and local long-range plan for our resources.

In the very first speech of your tour I think it is important to dramatize the whole issue by stating that you are leaving the Capital to make a nation-wide appeal for the preservation of our vital resources because this is America's last chance. That the next decade or so is in fact our last chance can be documented with a mass of bone chilling statistics -- these statistics and what they mean will paint a picture with a compelling force understandable to everyone. Raehel Carson's book on pesticides is a perfect example of the kind of impact that can be made with specifics. The situation is even worse in this country respecting water pollution, soil erosion, wildlife habitat destruction, vanishing open spaces, shortage of parks, etc.

As you well know, for more than a half century, conservationists have been writing, speaking and pleading for the preservation of our resources. Though the public is dimly aware that all around them, here and there, outdoor assets are disappearing, they really don't see the awful dimension of the catastrophe. The real failure has been in political leadership. This is a political issue to be settled at the political level, but strangely politicians seldom talk about it. Now for the first time in fifty years, conservationists have the President speaking for them. Since your voice will be heard, I think you should tell the whole story in your series of speeches.

The public should be told:

That there is no domestic issue more important to America in the long run than the conservation and proper use of our natural resources, including fresh water, clean air, tillable soil, forests, wilderness, habitat for wildlife, minerals and recreational assets.

That, in fact, our destruction of the landscape, the pollution of our air and waters, the overuse and abuse of our outdoor resources has proceeded at a pace in excess of any other culture in history.

That we need only look to the Middle East, China and India to see what happens to a culture and economy when it destroys its resources.

That the urgency of the issue right now is that the pace of our destruction has accelerated in the past 20 years and we have only another 10 or 15 years in which to take steps to conserve what is left.

Theodore Roosevelt said 50 years ago:

"...There is no question now before the nation of equal gravity with the question of the conservation of our natural resources."

There have been many other similar warnings before and since. But day in and day out, America has been too preoccupied with other problems to retain a sense of urgency about the crisis in our natural resources.

Recently, Brooks Atkinson, in the New York Times, reported: "No doubt, the history of American civilization could be written in terms of our changing attitudes toward nature. In 300 years we have passed through three significant stages: (1) indifference or hostility to nature; (2) romantic delight in nature, and now, (3) fear that man, the great predator, may destroy nature and civilization at the same time." In the same review he states: "Lois and Louis Darling conclude a study of the evolution and anatomy of birds ('Birds' is their title) with a chapter in the same somber key: 'We squander in a few years the fossil fuel, coal and oil, which are the accumulation of untold ages. We poison the water and the air.'

"In 'Face of North America,' Peter Farb makes a similar conclusion: 'The whole web of inter-relations developed in the wilderness over millions of years have been irretrievably lost.'

"If there is a fourth stage in American nature writing, it will portray a world short of food, cramped for space and bereft of beauty."

Americans in all walks of life are interested in natural resources. However, up to now there has not been any sustaining strong, central organization or leadership. Nevertheless, this interest is amazingly widespread. It cuts across political party lines, economic classes and geographical barriers.

The members of this vast interest group include all people in one way or another from ladies with a flower box in the window to the deer hunters with high powered rifles, the boaters, who range from kids with flat bottomed scows to the wealthy yachtsmen; family campers, whose numbers are growing rapidly; bird watchers; skin-divers; wilderness crusaders; farmers; soil conservationists; fishermen; insect collectors; foresters; just plain Sunday drivers, etc.

Most of these people have their own organization, some of them national as well as local, and they fight with a fury to advance and protect the phase of conservation in which they are interested. They will rally behind your leadership in this field in which most interested participants have the feeling that their personal interest is not sufficiently matched by official interest at the governmental level. Most of these groups also have their own national and local publications and their specialized news reporters and columnists, all of whom are hungry for materials and will be greatly stimulated by any demonstration of Presidential interest in their specialty.

The projected population growth gives one a frightening picture of the kind of pressure we will be putting against our resource base a few years from now. It is estimated that if our current population of 190 million persons continues to grow for only one century at its present rate, the population of the United States would be about 1 billion persons. This is equivalent to one third of the world's present inhabitants, and would be roughly equivalent to moving all of the population of Europe, Latin America and Africa into the territory of the 50 states. The whole populous eastern part of the United States is already in desperate circumstances for lack of recreational opportunity for the ordinary citizen, and the crisis grows worse each year.

I think it is most important that your series of speeches cover the widest possible range of resource preservation, beginning with a major speech in the east on the crisis in our densely populated areas. Then, it seems to me, all of the issues involved should be touched upon as you move across the country; hydroelectric power, recreation, reclamation, pollution of air and water, scenic beauty (or, scenic pollution), wilderness, seashore access, and the recreational issues including fishing, hunting, skiing, hiking, and camping In other words, I am suggesting that your series of speeches constitute a total presentation of the whole problem in both its broadest and its most specific aspects.

I also think that it is important to say something on the issue of comprehensive, national, state and local planning of long-term recreation needs, and the need for cooperation among all levels of government and all conservation groups. The Congress has already made a long step toward comprehensive planning by providing planning grants to states and localities for long-range planning, including recreational planning. It is an unfortunate fact that no state in the nation has a comprehensive long-range recreation plan, and will not have one until Wisconsin completes its plan with the use of Federal money, within one year from now. The same is true for all practical purposes of the Federal Government.

We have grown very rapidly. As the frontier pushed west, we failed to act every step of the way until it was too late -- recreation areas are gone, the lands are drained, the water is polluted. Only when our resources have been destroyed do we begin to worry and think and frantically search for some area to acquire and protect for the public to use.

I have rather rambled all over the lot in this letter, but the subject matter rambles that way too -- into everyone's life in a thousand ways.

In summary, I think it is important that you spell out the crisis in its broadest terms, and that every aspect of resource management and conservation be touched upon. This is necessary for public understanding. It is important also that every one of the hundreds of specialized magazines and newspapers be able to quote something from your speeches that is important to their field of interest and their readers.

Sincerely,

GAYLORD NELSON
U. S. Senator

Appendix 2

Introduction to "Environmental Agenda for Earth Day 1970"

As presented to the 91st Congress, January 19, 1970

In the nearly forty years since Franklin D. Roosevelt said in his first inaugural address that "this great nation will endure as it has endured, will revive and will prosper," our economy has soared to levels that no one in the 1930s could have imagined. In these past four decades we have become the wealthiest nation on earth by almost any measure of production and consumption.

As the economic boom and the postwar population explosion continued to break all records, a national legend developed: with science and technology as its tools the private enterprise system could accomplish anything.

We assumed that, if private enterprise could turn out more automobiles, airplanes and TV sets than all the rest of the world combined, somehow it could create a transportation system that would work. If we were the greatest builders in the world, we need not worry about our poor and about the planning and building of our cities. Private enterprise with enough technology and enough profit would manage that just fine.

In short, we assumed that if private enterprise could be such a spectacular success in the production of goods and services, it could do our social planning for us too, set our national priorities, shape our social system, and even establish our individual aspirations. In fact, I am sure most can recall the famous words of Charles Wilson back in the 1950s, when he said, "What's good for the country is good for General Motors, and vice versa."

In the 1960s the era of fantastic achievement marched on to levels unprecedented in the history of man. It was the decade when man walked on the moon; when medical magic transplanted the human heart; when the computer's mechanical wizardry became a part of daily life; and when, instead of "a chicken in every pot," the national aim seemed to be two cars in every garage, a summer home, a color television set, and a vacation home in Europe.

From the small farmers and small merchants of the last century, we had become the "consumer society," with science and technology as the New Testament and the gross national product as the Holy Grail. One might have thought we would have emerged triumphantly from the 1960s with a shout: "Bring on the next decade."

We have not. For, in addition to the other dramatic national and international events, the 1960s have produced another kind of "top of the decade" list. It has been a decade when the darkening cloud of pollution seriously began degrading the thin envelope of air surrounding the globe; when pesticides and unrestricted waste disposal threatened the productivity of all the oceans of the world; when virtually every lake, river and watershed in America began to show the distressing symptoms of being overloaded with polluting materials.

These pivotal events have begun to warn the nation of a disturbing new paradox: the mindless pursuit of quantity is destroying—not enhancing—the opportunity to achieve quality in our lives. In the words of the American balladeer, Pete Seeger, we have found ourselves "standing knee-deep in garbage, throwing rockets at the moon."

Cumulatively, "progress—American style" adds up each year to two hundred million tons of smoke and fumes, seven million junked cars, twenty million tons of paper, forty-eight billion cans, and twenty-eight billion bottles. It also means bulldozers gnawing away at the landscape to make room for more unplanned expansion, more leisure time but less open space in which to spend it, and so much reckless progress that we face even now a hostile environment.

As one measure of the rate of consumption that demands our resources and creates our vast wastes, it has been estimated that all the American children born in just one year would use up two hundred million pounds of steel, 9.1 billion gallons of gasoline, and twenty-five billion pounds of beef during their lifetimes.

To provide electricity for our air conditioners, the Kentucky hillside is strip-mined. To provide the gasoline for our automobiles, the ocean floor is drilled for oil. To provide the sites for our second homes, the shore of a pristine lake is subdivided. The unforeseen—or ignored—consequences of an urbanizing, affluent, mobile, more populous society have poisoned, scarred and polluted what once was a beautiful land "from sea to shining sea."

It is the laboring man, living in the shadows of the spewing smokestacks of industry, who feels the bite of the "disposable society." Or the commuter inching in spurts along an expressway. Or the housewife paying too much for products that began to fall apart too soon. Or the student watching the university building program destroy a community. Or the black man living alongside the noisy, polluted truck routes through the central city ghetto.

There is not merely irritation now with the environmental problems of daily life—there is a growing fear that what the scientists have been saying is all too true, that man is on the way to defining the terms of his own extinction.

Today it can be said that there is no clear air left in the United States. The last vestige of pure air was near Flagstaff, Ariz., but it disappeared six years ago. Today it can also be said that there is no river or lake in the country that has not been affected by the pervasive wastes of our society. On Lake Superior, the last clean Great Lake, a mining company is dumping 60,000 tons of iron ore process wastes a day directly into the lake.

Tomorrow? Responsible scientists have predicted that accelerating rates of air pollution could become so serious by the 1980s that many people may be forced on the worst days to wear breathing helmets to survive outdoors. It has also been predicted that in 20 years man will live in domed cities. Dr. S. Dillon Ripley, secretary of the Smithsonian Institution, believes that in twenty-five years somewhere between seventy-five and eighty percent of all the species of living animals will be extinct. Dr. Paul Ehrlich, eminent California ecologist, and many other scientists predict the end of the oceans as a productive resource within the next fifty years unless pollution is stopped. The United States provides an estimated one-third to one-half of the industrial pollution of the sea. It is especially ironic that, even as we pollute the sea, there is hope that its resources can be used to feed tens of millions of hungry people.

As in the Great Depression, America is again faced with a crisis that has to do with material things—but it is an entirely different sort of dilemma. In effect, America has bought environmental disaster on a national installment plan: buy affluence now and let future generations pay the price. Trading away the future is a high price to pay for an electric swizzle stick—or a car with greater horsepower. But then, the environmental consequences have never been included on the label.

It is a situation we have gotten into, not by design, but by default. Somehow, the environmental problems have mushroomed upon us from the blind side—although, again, the scientists knew decades ago that they were coming. What has been missing is the unity of purpose, forged out of the threat to our national health or security or prestige, that we so often seem to have found only during world war. But there is now, I think, a great awakening underway. We have begun to recognize that our security is again threatened—not from the outside, but from the inside—not by our enemies, but by ourselves. As Pogo quaintly put it, "We have met the enemy and they is us."

A Gallup poll taken for the National Wildlife Federation last year revealed that 51 percent of all persons interviewed were deeply disturbed about the grim tide of pollution. Growing student environmental concern is a striking new development. A freshman college student attitude poll, conducted last fall by the American Council on Education, found that 89.9 percent of all male freshmen believed the federal government should be more involved in the control of pollution. And a Gallup poll published in late December found that the control of air and water pollution is fast becoming a new student cause, with students placing the issue sixth on a list of areas where they felt changes must be made.

Other national and local polls, the rising citizen attendance at public hearings on polluters, the letters that are pouring into congressional offices—all indicate a vast new concern. As a dramatic indication of the degree the new citizen concern has reached Congress, a daily average of 150 constituent requests on environmental questions are coming into the Legislative Reference Service, the research arm of Congress, from members of Congress. This is a request rate second only to that for crime.

In the *Congressional Record,* the amount of environmental material inserted in the first six months of last year by senators and congressmen was exceeded only by material on the issue of Vietnam. Congress last year took the major initiative of appropriating $800 million in federal water pollution control funds—nearly four times the request of the present and previous administrations.

And environmentalists across the country have been heartened by reports that the president will devote major attention to the environmental crisis in his State of the Union message later this week. All conservationists applaud the President's interest and commitment. In short, I believe that today we are at a watershed in the history of the struggle in this country to save the quality of our environment.

With the massive new coalition of interests that is now forming, which is including the urbanite and the student, it is possible to wage war on our environment problems and win. In any such effort the continued commitment of millions of people is the most essential resource of all.

But, lest anyone be misled or caught unaware, this war would be lost before it is begun if we do not bring other massive resources to it as well. A victory will take decades and tens of billions of dollars. Just to control pollution, it will take $275 billion by the year 2000. Although that sounds like a lot of money, it will be spent over the next 30 years and is equivalent to the defense expenditure for the next four years.

More than money, restoring our environment and establishing quality on a par with quantity as a goal of American life will require a reshaping of our values, sweeping changes in the performance and goals of our institutions, national standards of quality for the goods we produce, a humanizing and redirection of our technology, and greatly increased attention to the problem of our expanding population.

Perhaps most of all, it will require on the part of the people a new assertion of environmental rights and the evolution of an ecological ethic of understanding and respect for the bonds that unite the species man with the natural systems of the planet.

The ecological ethic must be debated and evolved by individuals and institutions on the terms of man's interdependence with nature. Institutions such as our churches and universities could be of important assistance in providing increased understanding of these ethical considerations. Such an ethic, in recognizing the common heritage and concern of men of all nations, is the surest road to removing the mistrust and mutual suspicions that have always seemed to stand in the way of world peace.

American acceptance of the ecological ethic will involve nothing less than achieving a transition from the consumer society to a society of "new citizenship"—a society that concerns itself as much with the well-being of present and future generations as it does with bigness and abundance. It is an ethic whose yardstick for progress should be: is it good for people?

American college students—thousands of whom are now actively planning a teach-in on the crisis of the environment April 22 on hundreds of campuses—are in the forefront in expressing the terms on which we will need to meet this critical challenge. Students, scientists and many others are saying that we must reject any notion that progress means destroying Everglades National Park with massive airport development—or that it is progress to use the American public as an experimental laboratory for artificial sweeteners, food additives, or other products without understanding the "technological backlash" that may come from their unmeasured dangers—or that it is progress to fill hundreds of square miles of our bays and coastal wetlands, destroying natural habitat for thousands of species of fish and wildlife, polluting our waters, and in many other ways wreaking havoc with this fragile ecological system in the name of providing new space for industry, commerce, and subdivisions.

There is a great need, and growing support, for the introduction of new values in our society—where bigger is not necessarily better—where slower can be faster—and where less can be more. This attitude must be at the heart of the nationwide effort—an agenda for the 1970s—whereby this country puts gross national quality above gross national product.

Notes

3. Windows on the World

1. "The Amsterdam Declaration on Global Change," issued by the scientific communities of four international global change research programs: the International Geosphere-Biosphere Programme, the International Human Dimensions Programme on Global Environmental Change, the World Climate Research Programme, and the international biodiversity program DIVERSITAS. Prepared at the Global Change Open Science Conference in Amsterdam, Netherlands, July 13, 2001, <www.igbp.kva.se/> (accessed April 10, 2002).
2. Carl Sagan, *Billions and Billions: Thoughts on Life and Death at the Brink of the Millennium* (New York: Random House, 1997), 16–18.
3. Lester R. Brown, "Eradicating Hunger: A Growing Challenge," in *State of the World 2001*, by Lester R. Brown et al. (New York: Norton, 2001), 60–61.
4. Marla Cone, "Growth Slows as Population Hits 6 Billion," *Los Angeles Times*, October 12, 1999.
5. Population Reference Bureau, 2000 World Population Data Sheet.
6. Brown, "Eradicating Hunger," 45.
7. Jane Lubchenko, "Entering the Century of the Environment: A New Contract for Science," Presidential Address at the Annual Meeting of the American Association for the Advancement of Science, February 15, 1977; Food and Agriculture Organization of the United Nations (FAO), "The State of World Fisheries and Aquaculture, 2000," <www.fao.org/sof/sofia/> (accessed April 10, 2002).
8. Population and Consumption Task Force, President's Council on Sustainable Development, "Population and Consumption Task Force Report," 1996, <clinton2.nara.gov/PCSD/Publications/TF_Reports/pop-toc.html> (accessed April 10, 2002).
9. Ibid.
10. FAO, "State of World Fisheries and Aquaculture, 2000."

11. American Fisheries Society, "Marine, Estuarine, and Diadromous Fish Stocks at Risk of Extinction in North America (Exclusive of Pacific Salmonids)," *Fisheries Magazine,* November 2000.

12. Rodger Doyle, "Sprawling into the Third Millennium," *Scientific American,* March 2001.

13. Susan Milius, "Just How Much Do U.S. Roads Matter?" *Science News,* February 5, 2000, 95.

14. U.S. Department of Transportation, Bureau of Transportation Statistics, "National Transportation Statistics 2000," Report no. BTS01-01, April 2001; U.S. Environmental Protection Agency (EPA), "Automobiles and Ozone," January 1993 (updated July 1998), EPA 400-F-92-006, Fact Sheet OMS-4, <www.epa.gov/otaq/04-ozone.htm> (accessed April 10, 2002).

15. U.S. EPA, "Your Car and Clean Air: What YOU Can Do to Reduce Pollution," July 20, 1998, <www.epa.gov/otaq/18-youdo.htm> (accessed April 10, 2002).

16. Randolph E. Schmid, "Population Doubling: Twice as Many Americans by 2100," Associated Press, January 12, 2000.

17. David Pimentel, Mario Giampietro, and Sandra Bukkens, "An Optimum Population for North and Latin America," *Population and Environment* 20, no. 2 (November 1998): 125–48; David Pimentel and Mario Giampietro, "Food, Land, Population and the U.S. Economy," (Washington, D.C.: Carrying Capacity Network), <dieoff.org/page55.htm> (accessed April 10, 2002).

18. The Minnesota New Country School Frog Project, updated September 8, 1999, <www.mncs.k12.mn.us/html/projects/Frog/frog.html> (accessed April 9, 2002).

19. Northern Prairie Wildlife Research Center, North American Reporting Center for Amphibian Malformations, U.S. Geological Survey, Northern Prairie Wildlife Research Center home page, version March 27, 2002, <www.npwrc.usgs.gov/narcam> (accessed April 9, 2002).

20. Gary Casper, interview by Susan Campbell, March 21, 1998.

21. Katharine Q. Seelye, "Ending Logjam, U.S. Reaches Accord on Endangered Species," *New York Times,* August 30, 2001, A1.

22. World Conservation Union Species Survival Commission, "2000 IUCN Red List of Threatened Species" (Gland, Switzerland: World Conservation Union, 2000), tables 1q, 12b, <www.redlist.org> (accessed April 10, 2002).

23. John Tuxill, "Appreciating the Benefits of Plant Biodiversity," in *State of the World 1999,* by Lester R. Brown et al., Worldwatch Institute Books (New York: Norton, 1999), 97.

24. United Nations Environment Programme, *Global Environment Outlook 2000,* <www.grid.unep.ch/geo2000> (accessed February 28, 2002); Charlotte Schubert, "Life on the Edge: Will a Mass Extinction Usher in a World of Weeds and Pests?" *Science News* 160, September 15, 2001, 168–169.

25. "Biodiversity in the Next Millennium," a nationwide survey developed by the American Museum of Natural History and Louis Harris and Associ-

ates, Inc., in conjunction with the opening of the museum's new Hall of Bio-diversity, 1998; Schubert, "Life on the Edge," 170.

26. Susan Milius, "U.S. Biosurvey Reveals Worrisome Trends," *Science News* 156, September 25, 1999, 199; U.S. Geological Survey, "Status and Trends of the Nation's Biological Resources," September 20, 1999, <http://biology.usgs.gov/pubs/execsumm/> (accessed April 10, 2002); Rodger Doyle, "Plants at Risk in the U.S.," *Scientific American,* August 1997, 26.

27. "Mass Extinction of Freshwater Creatures Forecast," Environment News Service, October 4, 1999, <http://ens.lycos.com/ens/oct99/1999L-10-04-02.html> (accessed April 10, 2002).

28. U.S. Geological Survey, "Status and Trends."

29. William J. Broad, "Conservationists Write a Seafood Menu to Save Fish," *New York Times,* November 9, 1999, D3.

30. Mark Alpert, "Replumbing the Everglades," *Scientific American,* August 1999, 16.

31. Gary Casper, interview by Susan Campbell, March 21, 1998.

32. National Research Council, "Recounciling Observations of Global Temperature Change (Washington, D.C.: National Academy Press, 2000), 21.

33. Intergovernmental Panel on Climate Change, "Summary for Policymakers: A Report of Working Group 1 of the IPCC," 2001, 10.

34. United Nations Development Programme, *Human Development Report,* 1998.

35. Intergovernmental Panel on Climate Change, "Summary for Policymakers: A Report of Working Group 1 of the IPCC," 2001, 10.

36. National Research Council, Committee on the Science of Climate Change, "Climate Change Science: An Analysis of Some Key Questions" (Washington, D.C.: National Academy Press, 2001), 1.

37. Christopher Flavin, "Facing Up to the Risks of Climate Change," in *State of the World 1996,* by Lester R. Brown et al., Worldwatch Institute Books (New York: Norton, 1996), 22, 28.

38. Nina Leopold Bradley, telephone interview by Susan Campbell, August 2, 2001; Andre McCloskey, "Helping Species Cope with Change," *Audubon Naturalist News,* June 2001, <www.audubonnaturalist.org/feat0600.htm> (accessed April 10, 2002); Nina L. Bradley, A. Carl Leopold, John Ross, and Wellington Huffaker, "Phenological Changes Reflect Climate Change in Wisconsin," *Proceedings of the National Academy of Sciences* 96 (August 17, 1999) <www.pnas.org> (accessed April 10, 2002).

39. Thomas E. Graedel and Paul J. Crutzen, *Atmosphere, Climate and Change* (New York: Scientific American Library, 1995), 122.

40. Flavin, "Facing Up to the Risks," 26.

41. Ibid., 28; U.S. EPA, U.S. EPA's Global Warming Site, updated March 26, 2002, <www.epa.gov/globalwarming/visitorcenter/Coastal/index.html> (accessed April 10, 2002).

42. Richard Monastersky, "Acclimating to a Warmer World: With Some Cli-

mate Change Unavoidable, Researchers Focus on Adaptation," *Science News* 156, August 28, 1999, 137–38.

43. Ross Gelbspan, "Changing the Climate: The Global Warming Crisis," *Yes! A Journal of Positive Futures,* issue 12 (winter 1999): 16.

44. U.S. Department of Energy, Energy Information Administration, "Federal Financial Interventions and Subsidies in Energy Markets 1999," July 10, 2000, Report SR/OIAF/2000-02, <www.eia.doe.gov/oiaf/servicerpt/subsidy1/> (accessed April 10, 2002).

4. Vanishing Resources

1. William Mullen, "On the Record," *Chicago Tribune,* January 30, 2000, section 2, 3.

2. April Reese, "Bad Air Days," *E, The Environmental Magazine,* November/December 1999, 30.

3. Ibid.

4. U.S. Public Interest Research Group, "Danger in the Air: Unhealthy Smog Days in 1999," January 2000, <www.pirg.org/reports/enviro/smog/> (accessed April 10, 2002).

5. U.S. Environmental Protection Agency, Office of Air and Radiation, "1997 National Air Quality: Standards and Trends," December 1998, modified September 5, 2001, <www.epa.gov/oar/aqtrnd97/> (accessed April 10, 2002).

6. U.S. Environmental Protection Agency, Office of Air Quality Planning and Standards, "Latest Findings on National Air Quality: 2000 Status and Trends," September 2001, <www.epa.gov/oar/aqtrnd00/> (accessed June 13, 2002).

7. American Lung Association, news release, "Health Groups Cite Danger of Tiny Soot Particles, Call on EPA to Adopt Tough New Short-term Standard," July 23, 2001.

8. U.S. Environmental Protection Agency, Office of Mobile Sources, "Automobiles and Ozone," January 1993 (updated July 1998), EPA 400-F-92-006, Fact Sheet OMS-4, <www.epa.gov/otaq/04-ozone.htm> (accessed April 10, 2002).

9. R. Monastersky, "China's Air Pollution Chokes Crop Growth," *Science News,* March 27, 1999, 197, <www.sciencenews.org/sn_arc99/3_27_99/fob4.htm> (accessed April 10, 2002).

10. Don Hopey, "Regional Haze Means Visitors Can't Take in the Long View of America's Scenic Areas: A Poor View of the Vistas," *Pittsburgh Post-Gazette,* March 26, 2001; U.S. Environmental Protection Agency, "National Air Quality and Emissions Trends Report, 1997," December 10, 1998, <www.epa.gov/oar/aqtrnd97/trendsfs.html> (accessed April 10, 2002).

11. U.S. Environmental Protection Agency, Office of Mobile Sources, "Automobile Emissions: An Overview," August 1994 (updated July 1998), <www.epa.gov/omswww/05-autos.htm> (accessed April 10, 2002).

12. Ibid.; Howard W. Mielke, "Lead in the Inner Cities," *American Scientist* 87, no. 1 (January–February 1999), <www.sigmaxi.org/amsci/articles/99articles/mielke.html> (accessed April 10, 2002).

13. U.S. Department of Transportation, Bureau of Transportation Statistics, "National Transportation Statistics 2000," Report no. BTS01-01, April 2001; U.S. Environmental Protection Agency, "Automobile Emissions"; U.S. Environmental Protection Agency, "Automobiles and Ozone."

14. U.S. Environmental Protection Agency, "Your Car and Clean Air: What YOU Can Do to Reduce Pollution," August 1994 (updated July 1998), <www.epa.gov/otaq/18-youdo.htm> (accessed April 10, 2002).

15. U.S. Department of Energy, Office of Transportation Technologies, "Future U.S. Highway Energy Use: A Fifty-Year Perspective," May 3, 2001 (draft), <http://www.ott.doe.gov/pdfs/hwyfuture.pdf> (accessed April 10, 2002).

16. Congressional Research Service, Report to Congress, "Sport Utility Vehicles, Minivans and Light Trucks: An Overview of Fuel Economy and Emissions," updated September 16, 2001; Terril Yue Jones, "Passenger Cars Are Outsold by Light Trucks for First Time," *Los Angeles Times,* January 4, 2002, <www.latimes.com> (accessed April 10, 2002).

17. U.S. Department of Transportation, Bureau of Transportation Statistics, "National Transportation Statistics 2000," BTS01-01, table 1–9, April 2001.

18. National Research Council, Committee on Effectiveness and Impact of Corporate Average Fuel Economy (CAFE) Standards, Board on Energy and Environmental Systems, Transportation Research Board, *Effectiveness and Impact of Corporate Average Fuel Economy Standards* (Washington, D.C.: National Academy Press, 2001).

19. Carbon Dioxide Information Analysis Center, "Frequently Asked Global Change Questions," <http://cdiac.esd.ornl.gov/pns/faq.html> (accessed April 10, 2002).

20. Calculation assumes cars averaging 28 mpg, SUVs and light trucks averaging 19 mpg., <www.fueleconomy.gov> (accessed April 10, 2002).

21. National Research Council, *Effectiveness and Impact,* Executive Summary, 5-6.

22. Energy Information Administration, state petroleum profile sheets, August 2001.

23. U.S. Environmental Protection Agency, "National Water Quality Inventory: 1996 Report to Congress" and "National Water Quality Inventory: 1998 Report to Congress, <www.epa.gov> (accessed April 10, 2002).

24. Ibid.

25. Natural Resources Defense Council, "Testing the Waters 2001: A Guide to Water Quality at Vacation Beaches," 2001; NRDC, "Beach Closings Soar Following Better Testing and Reporting," press release, August 8, 200l; both at <www.nrdc.org/water/default.asp> (accessed April 10, 2002).

26. Steve Schultze and Marie Rohde, "Beach Closings Haven't Abated," *Milwaukee Journal Sentinel,* August 8, 2001, B-1; Steve Schultze, "Research Ties Gulls to Beach Pollution," *Milwaukee Journal Sentinel,* August 28, 2001, A-1.

27. Cameron Davis, telephone interview by Susan Campbell, August 2001.

28. U.S. Environmental Protection Agency, Office of Water, "Liquid Assets 2000," EPA-840-B-00-001, Executive Summary, May 2000, <www.epa.gov/ow/liquidassets/execsumm.html> (accessed April 10, 2002).

29. National Oceanic and Atmospheric Administration (NOAA), "State of the Coast Report," 1998, <state-of-coast.noaa.gov/> (accessed April 10, 2002).

30. T. E. Dahl, "Wetlands Losses in the United States, 1780s to 1980s," (Washington, D.C.: U.S. Department of the Interior, Fish and Wildlife Service, 1990), <http://www.npwrc.usgs.gov/resource/othrdata/wetloss/wetloss.htm> (accessed April 10, 2002).

31. David Zaring, "Federal Legislative Solutions to Agricultural Nonpoint Source Pollution," *ELR News and Analysis,* March 1996, 10136–37.

32. Marilyn Marchione, "Human Waste May Be Crypto Culprit: Study Suggests Source of Outbreak Wasn't Cattle," *Milwaukee Journal Sentinel,* October 18, 1997.

33. U.S. Environmental Protection Agency, "National Water Quality Inventory: 1996 Report to Congress."

34. National Research Council, Committee on the Toxicological Effects of Methylmercury, Board on Environmental Studies and Toxicology, *Toxicological Effects of Methylmercury,* (Washington, D.C.: National Academy Press, 2000), 322–27.

35. U.S. Centers for Disease Control, "Blood and Hair Mercury Levels in Young Children and Women of Childbearing Age—United States, 1999," *Morbidity and Mortality Weekly Report,* 50(08) (March 2, 2001): 140–43, <www.cdc.gov/mmwr/preview/mmwrhtml/mm5008a2.html/> (accessed March 23, 2002).

36. National Research Council, *Toxicological Effects of Methylmercury,* 2000, 1; U.S. Environmental Protection Agency, Office of Air and Radiation, "Mercury Study Report to Congress," December 1997, <www.epa.gov/airprogm/oar/mercury.html> (accessed April 10, 2002).

37. Joe Bower, "Water Wars," *Audubon* Magazine, March–April 2000, 16.

38. International Forum on Globalization, "Blue Gold: The Global Water Crisis and the Commodification of the World's Water Supply," June 1999, <www.ifg.org/bgsummary.html> (accessed April 10, 2002).

39. Sandra Postel, "Redesigning Irrigated Agriculture," in *State of the World 2000,* by Lester R. Brown et al., Worldwatch Institute Books (New York: Norton, 2000), 44.

40. Don Hinrichsen, "Solutions for a Water-Short World," *Population Reports* 26, no. 1 (September 1998).

41. Payal Sampat, "Deep Trouble: The Hidden Threat of Groundwater Pollution," Worldwatch Paper no. 154 (Washington, D.C.: Worldwatch Institute,

December 2000), 12, <www.worldwatch.org/pubs/paper/154excerpt.html> (accessed April 10, 2002).

42. Timothy Egan, "Near Vast Bodies of Water, Land Lies Parched," *New York Times,* August 12, 2001.

43. John Dodge, "Lack of Water Could Limit Growth," *Olympian* (Olympia, Washington), October 10, 2000.

44. W. M. Alley, T. E. Reilly, and O. L. Franke, "Sustainability of Ground-Water Resources," Denver, U.S. Geological Survey Circular 1186 (Washington, D.C.: Government Printing Office, 1999), 24–26, 65, <water.usgs.gov/pubs/circ/circ1186/> (accessed April 10, 2002).

45. Postel, "Redesigning Irrigated Agriculture," 42–43.

46. Susan Campbell, "Protectors Fear Great Lakes at Risk," *Green Bay (Wisconsin) Press-Gazette,* May 31, 1999, A1.

47. Michael O'Donnell and Jonathan Rademaekers, "Water Use Trends in the Southwestern United States 1950–1990," U.S. Geological Survey Web conference, July 7–25, 1997, <geochange.er.usgs.gov/sw/impacts/hydrology/water_use/> (accessed April 10, 2002); Jon Unruh and Diana Liverman, "Changing Water Use and Demand in the Southwest," U.S. Geological Survey Web conference, July 7–25, 1997, <geochange.er.usgs.gov/sw/impacts/society/water_demand> (accessed April 10, 2002); "Impact of Climate Change on Land Use in the Southwestern U.S," U.S. Geological Survey Web conference, July 7–25, 1997, <geochange.er.usgs.gov/sw/> (accessed April 10, 2002).

48. Postel, "Redesigning Irrigated Agriculture," 58.

49. John Wood and Gary Long, "Long Term World Oil Supply: A Resource Base/Production Path Analysis," Energy Information Administration, July 28, 2000, <www.eia.doe.gov/pub/oil_gas/petroleum/presentations/2000/long_term_supply/> (accessed April 10, 2002).

50. Colin J. Campbell and Jean H. Laherrère, "The End of Cheap Oil," *Scientific American,* March 1998, 78–83, <dieoff.org/page140.htm> (accessed April 10, 2002).

51. Richard B. Anderson, "A Not So Fond Farewell: Say Bye-Bye to Low Gas Prices," *Los Angeles Times,* April 12, 1999.

52. Energy Information Administration, "International Energy Outlook 2002," March 28, 2002, <www.eia.doe.gov/oiaf/ieo/> (accessed April 10, 2002).

53. Prepared statement of Mark Rubin, general manager for upstream sector, American Petroleum Institute, before the Senate Committee on Energy and Natural Resources, April 3, 2001.

54. Lester R. Brown, "Challenges of the New Century," in *State of the World 2000,* by Lester R. Brown et al., Worldwatch Institute Books, 17.

55. Mullen, "On the Record."

56. "Energy Star Success Story: St. Joseph Hospital," press release, <yosemite1.epa.gov/estar/business.nsf> (accessed April 10, 2002).

57. Interlaboratory Working Group on Energy-Efficient and Clean-Energy Technologies, "Scenarios for a Clean Energy Future," ORNL/CON-476 and

LBNL-44029, November 2000, <www.ornl.gov/ORNL/Energy_Eff/CEF.htm/> (accessed April 10, 2002).

58. Susan Campbell, "Is Crandon Mine Even Necessary?" *Green Bay Press-Gazette,* December 16, 1998, A5.

59. Michael Renner, *Working for the Environment: A Growing Source of Jobs* (Washington, D.C.: Worldwatch Institute, 2000); "Saving the Environment: A Jobs Engine for the 21st Century," Worldwatch news release, September 21, 2000, <www.worldwatch.org/alerts/000921.html> (accessed April 10, 2002).

60. Molly O'Meara, "Harnessing Information Technologies for the Environment," in *State of the World 2000,* by Lester R. Brown et al., Worldwatch Institute Books (New York: Norton, 2000), 128; Mark Dwortzan, "European Union Aims to Curb High Tech Pollution," *Environmental News Network,* September 13, 2000, <www.enn.com/features/2000/09/09132000/eurecycles_30541.asp> (accessed April 10, 2002).

61. U.S. Forest Service, "U.S. Forest Facts and Historical Trends," FS-696, March 2001, <www.fia.fs.fed.us/library/ForestFacts.pdf> (accessed April 10, 2002).

62. U.S. Forest Service, "A Gradual Unfolding of a National Purpose: A Natural Resource Agenda for the 21st Century," 1998, <www.fs.fed.us/news/agenda/sp30298.html> (accessed April 10, 2002).

63. Kathleen Wong, "A Pixel Worth 1,000 Words," *U.S. News and World Report,* July 19, 1999, 48–50.

64. U.S. Forest Service, North Carolina Division of Forestry, "Forest Inventory Analysis (FIA)," news release, Southern Environmental Law Center, July 16, 2001.

65. The Cascade Conservation Partnership (TCCP), "Protecting Old-Growth Forests," <www.ecosystem.org/tccp/protectforests> (accessed April 10, 2002).

66. H. Michael Anderson, "Reshaping National Forest Policy," *Issues in Science and Technology* (Magazine of the National Academies), Fall 1999, <bob.nap.edu/issues/16.1/anderson.htm> (accessed April 10, 2002).

67. U.S. Forest Service, "U.S. Forest Facts and Historical Trends," 10.

68. Alanna Mitchell, "Forests Face Global Extinction, Study Says," *Toronto Globe and Mail,* August 21, 2001.

69. Janet N. Abramovitz and Ashley T. Mattoon, "Recovering the Paper Landscape," in *State of the World 2000,* by Lester R. Brown et al., Worldwatch Institute Books (New York: Norton, 2000), 101–3.

70. Jack Ward Thomas, "What Now? From a Former Chief of the Forest Service," in *A Vision for the U.S. Forest Service: Goals for Its Next Century,* ed. Roger Sedjo (Washington, D.C.: Resources for the Future, 2000).

71. Iddo K. Wernick, Robert Herman, Shekhar Govind, and Jesse H. Ausubel, "Materialization and Dematerialization: Measures and Trends," *Daedalus* 125, no. 3 (summer 1996): 171–98; Abramovitz and Mattoon, "Recovering the Paper Landscape," 103.

72. Abramovitz and Mattoon, "Recovering the Paper Landscape," 102.

73. Robert C. Paehlke, "Environmental Values and Public Policy," *Environmental Policy in the 1990s: Reform or Reaction?* 3rd ed. (Washington, D.C.: Congressional Quarterly Press, 1997), 83.

74. Cat Lazaroff, "Noise at Grand Canyon Prompts Air Tour Restrictions," *Environmental News Network,* March 29, 2000, <http://ens.lycos.com/ens/mar2000/2000L-03-29-07.html> (accessed April 10, 2002); "Protecting Grand Canyon Quiet Affirmed by Court," Sierra Club press release, fall 1998; "FAA Cuts Back Grand Canyon Noise," *Environmental News Network,* July 28, 1999, <www.enn.com/enn-news-archive/1999/07072899/grandnoise.enn.doc_4571.asp> (accessed April 10, 2002); Donna Wiench, "National Parks Looking for Ways to Eliminate Unnatural Noise from Park Areas in Effort to Preserve Habitat," *National Public Radio,* October 5, 1999.

75. Michael Kraft and Norman Vig, "Environmental Policy from the 1970s to 2000: An Overview," in *Environmental Policy: New Directions for the Twenty-First Century,* ed. Vig and Kraft, 4th ed. (Washington, D.C.: Congressional Quarterly Press, 2000), 23.

76. Mike Dombeck, "Roadless Area Conservation: An Investment for Future Generations," January 5, 2001, <http://roadless.fs.fed.us/documents/rule/dombeck_stmt.htm> (accessed April 10, 2002).

77. "Yellowstone National Park and Snowmobiles Case Study," prepared by Egret Communications/ARA Consulting, July 2001.

78. Ted Williams, "Motorizing Public Land," *Audubon* Magazine, March–April 2000, 44–46.

79. Aldo Leopold, *Round River* (New York: Oxford University Press, 1993), 145–46.

5. An Invisible Threat

1. Agency for Toxic Substances and Disease Registry, "Public Health Statement for DDT, DDE, and DDD," May 1994, updated June 22, 2001, <www.atsdr.cdc.gov/toxprofiles/phs35.html> (accessed April 11, 2002).

2. Carolyn Shea, "Ask Audubon," *Audubon* Magazine, January-February 2000, 98; Anne Platt McGinn, "Phasing Out Persistent Organic Pollutants," *State of the World 2000,* by Lester R. Brown, Worldwatch Institute Books (New York: Norton, 2000), 83.

3. World Health Organization, malaria fact sheet, number 94, revised October 1998, <www.who.int/inf-fs/en/fact094.html> (accessed April 11, 2002).

4. McGinn, "Phasing Out Persistent Organic Pollutants," 83.

5. Rachel Carson, *Silent Spring* (New York: Houghton Mifflin, 1962).

6. Theo Colborn, Dianne Dumanoski, and John Peterson Meyers, *Our Stolen Future* (New York: Dutton, 1996), 136.

7. Arnold Aspelin and Arthur H. Grube, "Pesticides Industry Sales and Usage: 1996 and 1997 Market Estimates" (Washington, D.C.: U.S. Environmental

Protection Agency, 1999), <www.epa.gov/oppbead1/pestsales/97pestsales/> (accessed April 11, 2002).

8. National Research Council, *Pesticides in the Diets of Infants and Children* (Washington, D.C.: National Academy Press, 1993).

9. Pesticide Action Network North America, "Hooked on Poison: Pesticide Use in California, 1991–1998," May 3, 2000, <www.panna.org/resources/documents/hookedAvail.dv.html> (accessed April 11, 2002); James Liebman, "Rising Toxic Tide: Pesticide Use in California, 1991–1995," San Francisco: Pesticide Action Network North America and Californians for Pesticide Reform, 1997, <www.igc.apc.org/panna/risingtide/textoftide.html> (accessed April 11, 2002).

10. Pesticide Action Network North America, "Cancer-Causing Pesticide Use Rising in California: Report Shows Total Pesticide Use Remains Alarmingly High," "Hooked on Poison" news release, May 3, 2000, <www.panna.org/resources/documents/hookedMedia.dv.html> (accessed April 11, 2002).

11. McGinn, "Phasing Out Persistent Organic Pollutants," 87.

12. Ibid., 83.

13. U.S. Department of Health and Human Services, Public Health Service, Agency for Toxic Substances and Disease Registry, "Healthy Children—Toxic Environments: Acting on the Unique Vulnerability of Children Who Dwell Near Hazardous Waste Sites," Report of the Child Health Workgroup, presented to the Board of Scientific Counselors, April 28, 1997, <www.atsdr.cdc.gov/child/chw497.html> (accessed April 11, 2002).

14. National Research Council, *Pesticides in the Diets of Infants and Children*.

15. Ibid.

16. Consumers Union, "A Report Card for the EPA, Successes and Failures in Implementing the Food Quality Protection Act," February 2001, 3, <www.consumersunion.org/food/fqpa_info.htm> (accessed April 11, 2002).

17. U.S. Environmental Protection Agency, "EPA Acts to Reduce Children's Exposures to Two Older, Widely Used Pesticides," press release, August 2, 1999, <www.ecologic-ipm.com/epapr080299.html> (accessed April 11, 2002).

18. CNN.com.health, "EPA Bans Pesticide Dursban, Says Alternatives Available," June 8, 2000, <www.cnn.com/2000/HEALTH/06/08/dursban.ban.02/index.html> (accessed April 11, 2002).

19. Edward Groth III, Charles M. Benbrook, and Karen Lutz, "Update: Pesticides in Children's Foods. An Analysis of USDA PDP Data on Pesticide Residues," Yonkers, NY: Consumers Union, May 2000, <www.ecologic-ipm.com/PDP/Update_Childrens_Foods.pdf> (accessed April 11, 2002).

20. Edward Groth III, Charles M. Benbrook, and Karen Lutz, "Do You Know What You're Eating? An Analysis of U.S. Government Data on Pesticide Residues in Foods," Yonkers, NY: Consumers Union, February 1999, <www.consumersunion.org/food/do_you_know2.htm> (accessed April 11, 2002).

21. Physicians for Social Responsibility, "Environmental Endocrine Disruptors: What Health Care Providers Should Know," (Washington, D.C.: Physicians for Social Responsibility, 2001, <www.psr.org/enddisprimer.pdf> (accessed April 11, 2002).

22. U.S. Environmental Protection Agency, "TRI 1999 Data Release," Toxics Release Inventory (TRI) Program, updated February 2002, <www.epa.gov/triinter/tridata/tri99/index.htm> (accessed April 11, 2002).

23. U.S. Environmental Protection Agency, "National Water Quality Inventory: 1996 Report to Congress," <www.epa.gov/305b/96index.html> (accessed April 11, 2002).

24. Carson, *Silent Spring*, 179.

25. Theo Colborn, telephone interview by Susan Campbell, August 15, 1997.

26. Theo Colborn, telephone interview by Susan Campbell, October 10, 2001.

27. National Research Council, "Hormonally Active Agents in the Environment" (Washington, D.C.: National Academy Press, 1999).

28. Colborn, interview, October 10, 2001.

29. McGinn, "Phasing Out Persistent Organic Pollutants," 83–84.

30. Oral statement of U.S. EPA administrator Carol M. Browner before the New York State Assembly's Committee on Environmental Conservation, July 9, 1998.

6. Complacent Planet?

1. Gallup poll, March 5–7, 2001, April 3–9, 2000, <www.gallup.com/poll/topics/environment.asp> (accessed March 15, 2002).

2. Gallup poll, "Americans Are Environmentally Friendly, but Issue Not Seen as Urgent Problem," poll analyses by Lydia Saad and Riley E. Dunlap, *Gallup News Service,* April 17, 2000, <www.gallup.com/poll/releases/pr000417.asp> (accessed March 15, 2002).

3. Gallup poll, 1999.

4. Gallup poll, March 2001.

5. Christopher J. Bosso, "Environmental Groups and the New Political Landscape," *Environmental Policy: New Directions for the Twenty-first Century,* 4th ed. (Washington, D.C.: Congressional Quarterly Press, 2000), 64.

6. Roper poll, 2000.

7. Tibbett L. Speer, "Growing the Green Market," *American Demographics,* August 1997.

8. Commentary, "Big Orange Scores with Green Policy," *Atlanta Business Chronicle,* September 6, 1999.

9. Gallup poll, 2001.

10. Lydia Saad, "Environmental Concern Wanes in 1999 Earth Day Poll," *Gallup News Service,* April 22, 1999.

11. Gallup poll, January 2001; *Newsweek* poll, April 2000; Gallup poll, March 2001.

12. Speer, "Growing the Green Market."

13. Bruce Tonn and Carl Petrich, "Environmental Citizenship: Problems and Perspectives," National Center for Environmental Decision-making Research, Technical Report, NCEDR/97-16, October 1997, 91, <www.ncedr.org/pdf/citizenship.pdf> (accessed April 11, 2002); Harwood Group, "Yearning for Balance: Views of Americans on Consumption, Materialism, and the Environment" (Takoma Park, Md.: Merck Family Fund, 1995).

14. Quoted in Clayton Russell, "Wilderness Horizons Conference: Living and Learning in Sigurd Olson's Singing Wilderness," *Horizons,* Sigurd Olson Environmental Institute, winter 2000.

15. John Byrne Barry, "Victory: Henry Ford (Cough) Quits Smoking—Detroit Coalition Forces Hospital Incinerator Shutdown," Sierra Club, <www.sierraclub.org/planet/200110/victory.asp> (accessed April 11, 2002); Environmental Justice Resource Center. Clark Atlanta University, <www.ejrc.cau.edu/> (accessed April 11, 2002).

16. Paul Gorman, telephone interview by Susan Campbell, December 8, 1999.

17. Scott Darling, interview by Susan Campbell, "Building on Leopold's Legacy: Conservation for a New Century" conference, Madison, Wisconsin, October 6, 1999.

18. David A. Mazmanian and Michael E. Kraft, "The Three Epochs of the Environmental Movement," in *Toward Sustainable Communities,* ed. Mazmanian and Kraft (Cambridge: MIT Press, 1999), 15–24.

19. Michael Kraft and Norman Vig, "Environmental Policy from the 1970s to 2000: An Overview," *Environmental Policy: New Directions for the Twenty-first Century,* ed. Vig and Kraft, 4th ed. (Washington, D.C.: Congressional Quarterly Press, 2000), 14.

20. Sharon Beder, *Global Spin* (White River Junction, Vt.: Chelsea Green, 1998), 47.

21. Susan Campbell, "Conservationists Review Goals of Movement," *Green Bay (Wisconsin) Press-Gazette,* October 7, 1999, B3.

22. Michael Kraft, interview by Susan Campbell, Green Bay, Wisconsin, February 2000.

23. Times Mirror Center for the People and the Press, April 1994.

24. Beder, *Global Spin,* 177.

25. Ibid., 186.

26. Jeanne Chircop, "Using Dirty Words," *Mining Voice,* July/August 1999, 4.

27. Beder, *Global Spin,* 189.

28. Jim Motavalli, "The Learning Tree," *E Magazine,* September/October 1999, 33.

29. Michael Sanera and Jane S. Shaw, *Facts, Not Fear: A Parents' Guide to Teaching Children about the Environment,* 2nd ed. (Washington, D.C.: Regnery, 1999).

30. Richard Wilke, telephone interview by Susan Campbell, October 13, 2001.

31. National Environmental Education Training Foundation (NEETF)/Roper Starch poll, "Ninth Annual National Report Card: Lessons from the Environment," May 2001.

32. John Flicker, "The Audubon View: Taking on the Flat Earth Society," *Audubon* Magazine, November/December 1999, 8.

33. Beder, *Global Spin*, 27–28.

34. Carl Deal, *The Greenpeace Guide to Anti-Environmental Organizations* (Tucson: Odonian Press, 1993), 55–56, 70–71.

35. Norman Vig, "Presidential Leadership and the Environment," *Environmental Policy in the 1990s: Reform or Reaction?* 3rd ed. (Washington, D.C.: Congressional Quarterly Press, 1997), 112; Michael Kraft, "Environmental Policy in Congress: Revolution, Reform, or Gridlock?" in *Environmental Policy in the 1990s,* ed. Vig.

36. President's Council on Sustainable Development, Letter to President Clinton, May 5, 1999, <clinton2.nara.gov/PCSD/lettrep.html> (accessed April 11, 2002).

7. Achieving Sustainability

1. United Nations, "Report of the International Conference On Population and Development," Cairo, September 5–13, 1994, <www.iisd.ca/linkages/cairo.html> (accessed April 12, 2002).

2. Population and Consumption Task Force Report, President's Council on Sustainable Development, 1996, <clinton2.nara.gov/PCSD/Publications/TF_Reports/pop-toc.html> (accessed April 12, 2002).

3. National Environmental Education Advisory Council, "Report Assessing Environmental Education in the United States and the Implementation of the National Environmental Education Act of 1990—Report to Congress II," November 2000, <www.epa.gov/enviroed/neeac.html> (accessed April 12, 2002).

4. Aldo Leopold, *A Sand County Almanac: With Other Essays on Conservation from Round River,* (New York: Oxford University Press, 1966), 262.

5. Paul Gorman, telephone interview by Susan Campbell, December 8, 1999.

6. "Preserving and Cherishing the Earth: An Appeal for Joint Commitment in Science and Religion." The document was organized by Dr. Carl Sagan and presented at the Moscow meeting of the Global Forum of Spiritual and Parliamentary Leaders, January 15–19, 1990. A copy of the letter can be found at National Religious Partnership for the Environment, "An Open Letter to the Religious Community, January 1990," <www.nrpe.org/history.html> (accessed April 12, 2002).

7. Carl Sagan, *Billions and Billions: Thoughts on Life and Death at the Brink of the Millennium* (New York: Random House, 1997), 146.

Acknowledgments

The authors would like to thank a number of people who gave freely of their time or helped in other ways to bring this book to publication.

Thank you to Jim Wiersma and others in the scientific and environmental community who took the time to proof early drafts of the manuscript for technical accuracy, and to Tom Content for his many editing contributions and insightful suggestions. We'd also like to thank Chuck Lacasse for his lead role in preparing the graphics, and Sally Jo Bongle and Kelley Rudolph for their contributions.

Thanks to Susanne Breckenridge at the University of Wisconsin Press for her flexibility and careful attention to the manuscript, and to Raphael Kadushin, Sheila Moermond, and Andrea Christofferson, among others. Thanks also to Jane Curran, for her thoughtful and thorough editing—and her enthusiasm.

The authors would like to acknowledge those who provided technical support during various stages of preparing the manuscript, among them Michael Westura and Joanne Lamb. Thanks also to Sam Wozniak for his work on the Web site, which we hope will be a continuing source of information for those interested in learning more about the ever-evolving field of environmental science and policy.

Finally, a special thank you to Art Kaftan, for making the introductions that made this book possible.

Index

193